THE ASTRONOMY

OF

MILTON'S 'PARADISE LOST'

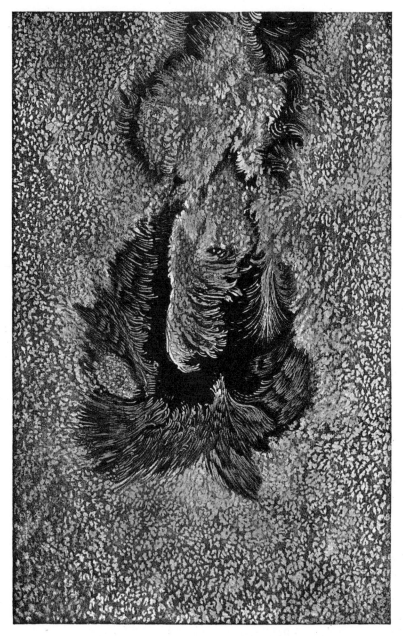

A TYPICAL SUN-SPOT

THE ASTRONOMY

OF

MILTON'S 'PARADISE LOST'

BY

THOMAS N. ORCHARD, M.D.

MEMBER OF THE BRITISH ASTRONOMICAL ASSOCIATION

These are thy glorious works, Parent of good,
Almighty! thine this universal frame,
Thus wondrous fair: Thyself how wondrous then!
Unspeakable.

HASKELL HOUSE
Publishers of Scholarly Books
NEW YORK
1966

CONTENTS

ILLUSTRATIONS

PLATES

IN TEXT

THE ASTRONOMY

OF

MILTON'S 'PARADISE LOST'

———◦◦———

CHAPTER I

A SHORT HISTORICAL SKETCH OF ASTRONOMY

ASTRONOMY is the oldest and most sublime of all
the sciences. To a contemplative observer of the
heavens, the number and brilliancy of the stars, the
lustre of the planets, the silvery aspect of the Moon,
with her ever-changing phases, together with the
order, the harmony, and unison pervading them all,
create in his mind thoughts of wonder and admiration. Occupying the abyss of space indistinguishable from infinity, the starry heavens in grandeur
and magnificence surpass the loftiest conceptions
of the human mind ; for, at a distance beyond the
range of ordinary vision, the telescope reveals
clusters, systems, galaxies, universes of stars—
suns—the innumerable host of heaven, each shining
with a splendour comparable with that of our Sun,
and, in all likelihood, fulfilling in a similar manner
the same beneficent purposes.

The time when man began to study the stars is

B

lost in the antiquity of prehistoric ages. The ancient inhabitants of the Earth regarded the heavenly bodies with veneration and awe, erected temples in their honour, and worshipped them as deities. Historical records of astronomy carry us back several thousand years. During the greater part of this time, and until a comparatively recent period, astronomy was associated with astrology—a science which originated from a desire on the part of mankind to penetrate the future, and which was based upon the supposed influence of the heavenly bodies upon human and terrestrial affairs. It was natural to imagine that the overruling power which governed and directed the course of sublunary events resided in the heavens, and that its decrees might be understood by watching the movements of the heavenly bodies under its control. It was, therefore, believed that by observing the configuration of the planets and the positions of the constellations at the instant of the birth of an individual, his horoscope, or destiny, could be foretold; and that by making observations of a somewhat similar nature the occurrence of events of public importance could be predicted. When, however, the laws which govern the motions of the heavenly bodies became better known, and especially after the discovery of the great law of gravitation, astrology ceased to be a belief, though for long after it retained its power over the imagination, and was often alluded to in the writings of poets and other authors.

In the early dawn of astronomical science, the theories upheld with regard to the structure of the heavens were of a simple and primitive nature, and might even be described as grotesque. This need occasion no surprise when we consider the difficulties with which ancient astronomers had to contend in their endeavours to reduce to order and harmony the complicated motions of the orbs which they beheld circling around them.

The grouping of the stars into constellations having fanciful names, derived from fable or ancient mythology, occurred at a very early period, and though devoid of any methodical arrangement, is yet sufficiently well-defined to serve the purposes of modern astronomers. Several of the ancient nations of the earth, including the Chaldeans, Egyptians, Hindus, and Chinese, claim to have been the earliest astronomers. Chinese records of astronomy reveal an antiquity of near 3,000 years B.C., but they contain no evidence that their authors possessed any scientific knowledge, and they merely record the occurrence of solar eclipses and the appearances of comets.

It is not known when astronomy was first studied by the Egyptians; but what astronomical information they have handed down is not of a very intelligible kind, nor have they left behind any data that can be relied upon. The Great Pyramid, judging from the exactness with which it faces the cardinal points, must have been designed by persons who possessed a good knowledge of astronomy, and it was probably made use of for observational purposes.

It is now generally admitted that correct astronomical observations were first made on the plains of Chaldea, records of eclipses having been discovered in Chaldean cities which date back 2,234 years B.C. The Chaldeans were true astronomers : they made correct observations of the risings and settings of the heavenly bodies; and the exact orientation of their temples and public buildings indicates the precision with which they observed the positions of celestial objects. They invented the zodiac and gnomon, made use of several kinds of dials, notified eclipses, and divided the day into twenty-four hours.

To the Greeks belongs the credit of having first studied astronomy in a regular and systematic manner. THALES (640 B.C.) was one of the earliest of Greek astronomers, and may be regarded as the founder of the science among that people. He was born at Miletus, and afterwards repaired to Egypt for the purpose of study. On his return to Greece he founded the Ionian school, and taught the sphericity of the Earth, the obliquity of the ecliptic, and the true causes of eclipses of the Sun and Moon. He also directed the attention of mariners to the superiority of the Lesser Bear, as a guide for the navigation of vessels, as compared with the Great Bear, by which constellation they usually steered. Thales believed the Earth to be the centre of the universe, and that the stars were composed of fire ;

he also predicted the occurrence of a great solar eclipse.

Thales had for his successors Anaximander, Anaximenes, and Anaxagoras, who taught the doctrines of the Ionian school.

The next great astronomer that we read of is PYTHAGORAS, who was born at Samos 590 B.C. He studied under Thales, and afterwards visited Egypt and India, in order that he might make himself familiar with the scientific theories adopted by those nations. On his return to Europe he founded his school in Italy, and taught in a more extended form the doctrines of the Ionian school. In his speculations with regard to the structure of the universe he propounded the theory (though the reasons by which he sustained it were fanciful) that the Sun is the centre of the planetary system, and that the Earth revolves round him. This theory—the accuracy of which has since been confirmed—received but little attention from his successors, and it sank into oblivion until the time of Copernicus, by whom it was revived. Pythagoras discovered that the Morning and Evening Stars are one and the same planet.

Among the famous astronomers who lived about this period we find recorded the names of Meton, who introduced the Metonic cycle into Greece and erected the first sundial at Athens ; Eudoxus, who persuaded the Greeks to adopt the year of $365\frac{1}{4}$ days ; and Nicetas, who taught that the Earth completed a daily revolution on her axis.

The Alexandrian school, which flourished for three centuries prior to the Christian era, produced men of eminence whose discoveries and investigations, when arranged and classified, enabled astronomy to be regarded as a true theoretical science. The positions of the fixed stars and the paths of the planets were determined with greater accuracy, and irregularities of the motions of the Sun and Moon were investigated with greater precision. Attempts were made to ascertain the distance of the Sun from the Earth, and also the dimensions of the terrestrial sphere. The obliquity of the ecliptic was accurately determined, and an arc of the meridian was measured between Syene and Alexandria. The names of Aristarchus, Eratosthenes, Aristyllus, Timocharis, and Autolycus, are familiarly known in association with the advancement of the astronomy of this period.

We now reach the name of HIPPARCHUS of Bithynia (140 B.C.), the most illustrious astronomer of antiquity, who did much to raise astronomy to the position of a true science, and who has also left behind him ample evidence of his genius ' as a mathematician, an observer, and a theorist.' We are indebted to him for the earliest star catalogue, in which he included 1,081 stars. He discovered the Precession of the Equinoxes, and determined the motions of the Sun and Moon, and also the length of the year, with greater precision than any of his predecessors. He invented the sciences of plane and spherical trigonometry, and was the first to use right ascensions and declinations.

The next astronomer of eminence after Hipparchus was PTOLEMY (130 A.D.), who resided at Alexandria. He was skilled as a mathematician and geographer, and also excelled as a musician. His chief discovery was an irregularity of the lunar motion, called the ' evection.' He was also the first to observe the effect of the refraction of light in causing the apparent displacement of a heavenly body from its true position. Ptolemy devoted much of his time to extending and improving the theories of Hipparchus, and compiled a great treatise, called the ' Almagest,' which contains nearly all the knowledge we possess of ancient astronomy. Ptolemy's name is, however, most widely known in association with what is called the Ptolemaic theory. This system, which originated long before his time, but of which he was one of the ablest expounders, was an attempt to establish on a scientific basis the conclusions and results arrived at by early astronomers who studied and observed the motions of the heavenly bodies. Ptolemy regarded the Earth as the immovable centre of the universe, round which the Sun, Moon, planets, and the entire heavens completed a daily revolution in twenty-four hours. After the death of Ptolemy no worthy successor was found to occupy his place, the study of astronomy began to decline among the Greeks, and after a time it ceased to be cultivated by that people.

The Arabs next took up the study of astronomy, which they prosecuted most assiduously for a period of four centuries. Their labours were, how-

ever, confined chiefly to observational work, in
which they excelled; unlike their predecessors,
they paid but little attention to speculative theories
—indeed, they regarded with such veneration the
opinions held by the Greeks, that they did not feel
disposed to question the accuracy of their doctrines.
The most eminent astronomer among the Arabs was
ALBATEGNIUS (680 A.D.). He corrected the Greek
observations, and made several discoveries which
testified to his abilities as an observer. IBN YUNIS
and ABUL WEFU were Arab astronomers who earned
a high reputation on account of the number and accuracy of their observations. In Persia, a descendant
of the famous Genghis Khan erected an observatory,
where astronomical observations were systematically
made. Omar, a Persian astronomer, suggested a
reformation of the calendar which, if it had been
adopted, would have insured greater accuracy than
can be attained by the Gregorian style now in use.
In 1433, Ulugh Beg, who resided at Samarcand,
made many observations, and constructed a star
catalogue of greater exactness than was known to
exist prior to his time. The Arabs may be regarded
as having been the custodians of astronomy until
the time of its revival in another quarter of the
Globe.

After the lapse of many centuries, astronomy
was introduced into Western Europe in 1220, and
from that date to the present time its career has
been one of triumphant progress. In 1230, a translation of Ptolemy's 'Almagest' from Arabic into

Latin was accomplished by order of the German Emperor, Frederick II. ; and in 1252 Alphonso X., King of Castile, himself a zealous patron of astronomy, caused a new set of astronomical tables to be constructed at his own expense, which, in honour of his Majesty, were called the 'Alphonsine Tables.' Purbach and Regiomontanus, two German astronomers of distinguished reputation, and Waltherus, a man of considerable renown, made many important observations in the fifteenth century.

The most eminent astronomer who lived during the latter part of this century was Copernicus. NICOLAS COPERNICUS was born February 19, 1473, at Thorn, a small town situated on the Vistula, which formed the boundary between the kingdoms of Prussia and Poland. His father was a Polish subject, and his mother of German extraction. Having lost his parents early in life, he was educated under the supervision of his uncle Lucas, Bishop of Ermland. Copernicus attended a school at Thorn, and afterwards entered the University of Cracow, in 1491, where he devoted four years to the study of mathematics and science. On leaving Cracow he attached himself to the University of Bologna as a student of canon law, and attended a course of lectures on astronomy given by Novarra. In the ensuing year he was appointed canon of Frauenburg, the cathedral city of the Diocese of Ermland, situated on the shores of the Frisches Haff. In the year 1500 he was at Rome, where he lectured on mathematics and astronomy. He next spent a few years at the University of Padua, where, besides applying himself to mathematics and astronomy, he studied medicine and obtained a degree. In 1505 Copernicus returned to his native country, and was appointed medical attendant to his uncle, the Bishop of Ermland, with whom he resided in the stately castle of Heilsberg, situated at a distance of forty-six miles from Frauenburg. Copernicus lived with his uncle from 1507 till 1512, and during that time prosecuted his astronomical studies, and undertook, besides, many arduous duties associated with the administration of the diocese ; these he faithfully discharged until the death of the Bishop, which occurred in 1512. After the death of his uncle he took up his residence at Frauenburg, where he occupied his time in meditating on his new astronomy and undertaking various duties of a public character, which he fulfilled with credit and distinction. In 1523 he was appointed Administrator-General of the diocese. Though a canon of Frauenburg, Copernicus never became a priest.

After many years of profound meditation and thought, Copernicus, in a treatise entitled 'De Revolutionibus Orbium Celestium,' propounded a new theory, or, more correctly speaking, revived the ancient Pythagorean system of the universe. This great work, which he dedicated to Pope Paul III., was completed in 1530 ; but he could not be prevailed upon to have it published until 1543, the year in which he died. In 1542 Copernicus had an apoplectic seizure, followed by paralysis and a gradual decay of his mental and vital powers. His book was printed at Nuremberg, and the first copy arrived at Frauenburg on May 24, 1543, in time to be touched by the hands of the dying man, who in a few hours after expired. The house in which Copernicus lived at Allenstein is still in existence, and in the walls of his chamber are visible the perforations which he made for the purpose of observing the stars cross the meridian.

Copernicus was the means of creating an entire revolution in the science of astronomy, by transferring the centre of our system from the Earth to the Sun. He accounted for the alternation of day and night by the rotation of the Earth on her axis, and for the vicissitudes of the seasons by her revolution round the Sun. He devoted the greater part of his life to meditating on this theory, and adduced several weighty reasons in its support. Copernicus could not help perceiving the complications and entanglements by which the Ptolemaic system of the universe was surrounded, and which compared unfavourably with the simple and orderly manner in which other natural phenomena presented themselves to his observation. By perceiving that Mars when in opposition was not much inferior in lustre to Jupiter, and when in conjunction resembled a star of the second magnitude, he arrived at the conclusion that the Earth could not be the centre of the planet's motion. Having discovered in some ancient manuscripts a theory, ascribed to the Egyptians, that Mercury

and Venus revolved round the Sun, whilst they
accompanied the orb in his revolution round the
Earth, Copernicus was able to perceive that this
afforded him a means of explaining the alternate
appearance of those planets on each side of the
Sun.   The varied aspects of the superior planets,
when observed in different parts of their orbits, also
led him to conclude that the Earth was not the
central body round which they accomplished their
revolutions. As a combined result of his observation
and reasoning Copernicus propounded the theory
that the Sun is the centre of our system, and that
all the planets, including the Earth, revolve in orbits
around him.   This, which is called the Copernican
system, is now regarded as, and has been proved to
be, the true theory of the solar system.

TYCHO BRAHÉ was a celebrated Danish astronomer, who earned a deservedly high reputation on
account of the number and accuracy of his astronomical observations and calculations. The various
astronomical tables that were in use in his time
contained many inaccuracies, and it became necessary that they should be reconstructed upon a
more correct basis. Tycho possessed the practical
skill required for this kind of work.

He was born December 14, 1546, at Knudstorp,
near Helsingborg. His father, Otto Brahé, traced
his descent from a Swedish family of noble birth.
At the age of thirteen Tycho was sent to the University of Copenhagen, where it was intended he
should prepare himself for the study of the law.

The prediction of a great solar eclipse, which was to happen on August 21, 1560, caused much public excitement in Denmark, for in those days such phenomena were regarded as portending the occurrence of events of national importance. Tycho looked forward with great eagerness to the time of the eclipse. He watched its progress with intense interest, and when he perceived all the details of the phenomenon occur exactly as they were predicted, he resolved to pursue the study of a science by which, as was then believed, the occurrence of future events could be foretold. From Copenhagen Tycho Brahé was sent to Leipsic to study jurisprudence, but astronomy absorbed all his thoughts. He spent his pocket-money in purchasing astronomical books, and, when his tutor had retired to sleep, he occupied his time night after night in watching the stars and making himself familiar with their courses. He followed the planets in their direct and retrograde movements, and with the aid of a small globe and pair of compasses was able by means of his own calculations to detect serious discrepancies in the Alphonsine and Prutenic tables. In order to make himself more proficient in calculating astronomical tables he studied arithmetic and geometry, and learned mathematics without the aid of a master. Having remained at Leipsic for three years, during which time he paid far more attention to the study of astronomy than to that of law, he returned to his native country in consequence of the death of an uncle, who bequeathed him a considerable estate. In Denmark he continued to prosecute his astronomical studies, and incurred the displeasure of his friends, who blamed him for neglecting his intended profession and wasting his time on astronomy, which they regarded as useless and unprofitable.

Not caring to remain among his relatives, Tycho Brahé returned to Germany, and arrived at Wittenberg in 1566. Whilst residing here he had an altercation with a Danish gentleman over some question in mathematics. The quarrel led to a duel with swords, which terminated rather unfortunately for Tycho, who had a portion of his nose cut off. This loss he repaired by ingeniously contriving one of gold, silver, and wax, which was said to bear a good resemblance to the original. From Wittenberg Tycho proceeded to Augsburg, where he resided for two years. Here he made the acquaintance of several men distinguished for their learning and their love of astronomy. During his stay at Augsburg he constructed a quadrant of fourteen cubits radius, on which were indicated the single minutes of a degree; he made many valuable observations with this instrument, which he used in combination with a large sextant.

In 1571 Tycho returned to Denmark, where his fame as an astronomer had preceded him, and was the means of procuring for him a hearty welcome from his relatives and friends. In 1572, when returning one night from his laboratory—for Tycho studied alchemy as well as astronomy—he beheld what appeared to be a new and brilliant star in the constellation Cassiopeia, which was situated overhead. He directed the attention of his companions to this wonderful object, and all declared that they had never observed such a star before. On the following night he measured its distance from the nearest stars in the constellation, and arrived at the conclusion that it was a fixed star, and beyond our system.

This remarkable object remained visible for sixteen months, and when at its brightest rivalled Sirius. At first it was of a brilliant white colour, but as it diminished in size it became yellow ; it next changed to a red colour, resembling Aldebaran ; afterwards it appeared like Saturn, and as it grew smaller it decreased in brightness, until it finally became invisible. In 1573 Tycho Brahé married a peasant-girl from the village of Knudstorp. This imprudent act roused the resentment of his relatives, who, being of noble birth, were indignant that he should have contracted such an alliance. The bitterness and mutual ill-feeling created by this affair became so intense that the King of Denmark deemed it advisable to endeavour to bring about a reconciliation.

After this Tycho returned to Germany, and visited several cities before deciding where he should take up his permanent residence.

His fame as an astronomer was now so great that he was received with distinction wherever he went, and on the occasion of a visit to Hesse-Cassel he spent a few pleasant days with William, Land-grave of Hesse, who was himself skilled in astro-nomy.

Frederick II., King of Denmark, having recog-nised Tycho Brahé's great merits as an astronomer, and not wishing that his fame should add lustre to a foreign Court, expressed a desire that he should return to his native country, and as an inducement offered him a life interest in the island of Huen, in the Sound, where he undertook to erect and equip an observatory at his own expense; the King also promised to bestow upon him a pension, and grant him other emoluments besides.

Tycho gladly accepted this generous offer, and during the construction of the observatory occupied his time in making a magnificent collection of instruments and appliances adapted for observa-tional purposes. This handsome edifice, upon which the King of Denmark expended a sum of 20,000*l*., was called 'Uranienburg' (' The Citadel of the Heavens '). Here Tycho resided for a period of twenty years, during which time he pursued his astronomical labours with untiring energy and zeal, and made a large number of observations and calculations of much superior accuracy to any that existed previously, which were afterwards of great service to his successors. During his long residence at Huen, Tycho was visited by many distinguished persons, who were attracted to his island home by his fame and the magnificence of his observatory. Among them was James VI. of Scotland, who, whilst journeying to the Court of Denmark on the occasion of his marriage to a Danish princess, paid Tycho a visit, and enjoyed his hospitality for a week. The King was delighted with all that he saw, and on his departure presented Tycho with a handsome donation, and at his request composed some Latin verses, in which he eulogised his host and praised his observatory.

The island of Huen is situated about six miles from the coast of Zealand, and fourteen from Copenhagen. It has a circumference of six miles, and consists chiefly of an elevated plateau, in the centre of which Tycho erected his observatory, the site of which is now marked by two pits and a few mounds of earth—all that remains of Uranienburg. All went well with Tycho Brahé during the lifetime of his noble patron ; but in 1588 Frederick II. died, and was succeeded by his son, a youth eleven years of age.

The Danish nobles had long been jealous of Tycho's fame and reputation, and on the death of the King an opportunity was afforded them of intriguing with the object of accomplishing his downfall. Several false accusations were brought against him, and the Court party made the impoverished state of the Treasury an excuse for depriving him of his pension and emoluments granted by the late King.

Tycho was no longer able to bear the expense of maintaining his establishment at Huen, and fearing that he might be deprived of the island itself, he took a house in Copenhagen, to which he removed all his smaller instruments.

During his residence in the capital he was subjected to annoyance and persecution. An order was issued in the King's name preventing him from carrying on his chemical experiments, and he besides suffered the indignity of a personal assault. Tycho Brahé resolved to quit his ungrateful country and seek a home in some foreign land, where he should be permitted to pursue his studies unmolested and live in quietness and peace. He accordingly removed from the island of Huen all his instruments and appliances that were of a portable nature, and packed them on board a vessel which he hired for the purpose of transport, and, having embarked with his family, his servants, and some of his pupils and assistants, ' this interesting barque, freighted with the glory of Denmark,' set sail from Copenhagen about the end of 1597, and having crossed the Baltic in safety, arrived at Rostock, where Tycho found some old friends waiting to receive him. He was now in doubt as to where he should find a home, when the Austrian Emperor Rudolph, himself a liberal patron of science and the fine arts, having heard of Tycho Brahé's misfortunes, sent him an invitation to take up his abode in his dominions, and promised that he should be treated in a manner worthy of his reputation and fame.

Tycho resolved to accept the Emperor's kind invitation, and in the spring of 1599 arrived at

Prague, where he found a handsome residence prepared for his reception.

He was received by the Emperor in a most cordial manner and treated with the greatest kindness. An annual pension of three thousand crowns was settled upon him for life, and he was to have his choice of several residences belonging to his Majesty, where he might reside and erect a new observatory. From among these he selected the Castle of Benach, in Bohemia, which was situated on an elevated plateau and commanded a wide view of the horizon.

During his residence at Benach Tycho received a visit from Kepler, who stayed with him for several months in order that he might carry out some astronomical observations. In the following year Kepler returned, and took up his permanent residence with Tycho, having been appointed assistant in his observatory, a post which, at Tycho's request, was conferred upon him by the Emperor.

Tycho Brahé soon discovered that his ignorance of the language and unfamiliarity with the customs of the people caused him much inconvenience. He therefore asked permission from the Emperor to be allowed to remove to Prague. This request was readily granted, and a suitable residence was provided for him in the city.

In the meantime his family, his large instruments, and other property, having arrived at Prague, Tycho was soon comfortably settled in his new home.

c 2

Though Tycho Brahé continued his astronomical observations, yet he could not help feeling that he lived among a strange people ; nor did the remembrance of his sufferings and the cruel treatment he received at the hands of his fellow-countrymen subdue the affection which he cherished towards his native land. Pondering over the past, he became despondent and low-spirited ; a morbid imagination caused him to brood over small troubles, and gloomy, melancholy thoughts possessed his mind—symptoms which seemed to presage the approach of some serious malady. One evening, when visiting at the house of a friend, he was seized with a painful illness, to which he succumbed in less than a fortnight. He died at Prague on October 24, 1601, when in his fifty-fifth year.

The Emperor Rudolph, when informed of Tycho Brahé's death, expressed his deep regret, and commanded that he should be interred in the principal church in the city, and that his obsequies should be celebrated with every mark of honour and respect.

Tycho Brahé stands out as the most romantic and prominent figure in the history of astronomy. His independence of character, his ardent attachments, his strong hatreds, and his love of splendour, are characteristics which distinguish him from all other men of his age. This remarkable man was an astronomer, astrologer, and alchemist; but in his latter years he renounced astrology, and believed that the stars exercised no influence over the destinies of mankind.

As a practical astronomer, Tycho Brahé has not been excelled by any other observer of the heavens. The magnificence of his observatory at Huen, upon the equipment and embellishment of which it is stated he expended a ton of gold ; the splendour and variety of his instruments, and his ingenuity in inventing new ones, would alone have made him famous. But it was by the skill and assiduity with which he carried out his numerous and important observations that he has earned for himself a position of the most honourable distinction among astronomers. In his investigation of the Lunar theory Tycho Brahé discovered the Moon's *annual equation*, a yearly effect produced by the Sun's disturbing force as the Earth approaches or recedes from him in her orbit. He also discovered another inequality in the Moon's motion, called the *variation*. He determined with greater exactness astronomical refractions from an altitude of 45° downwards to the horizon, and constructed a catalogue of 777 stars. He also made a vast number of observations on planets, which formed the basis of the 'Rudolphine Tables,' and were of invaluable assistance to Kepler in his investigation of the laws relating to planetary motion.

Tycho Brahé declined to accept the Copernican theory, and devised a system of his own, which he called the 'Tychonic.' By this arrangement the Earth remained stationary, whilst all the planets revolved round the Sun, who in his turn completed a daily revolution round the Earth. All the phenomena associated with the motions of those bodies could be explained by means of this system; but it did not receive much support, and after the Copernican theory became better understood it was given up, and heard of no more.

We now arrive at the name of KEPLER, one of the very greatest of astronomers, and a man of remarkable genius, who was the first to discover the real nature of the paths pursued by the Earth and planets in their revolution round the Sun. After seventeen years of close observation, he announced that those bodies travelled round the Sun in elliptical or oval orbits, and not in circular paths, as was believed by Copernicus. In his investigation of the laws which govern the motions of the planets he formulated those famous theorems known as 'Kepler's Laws,' which will endure for all time as a proof of his sagacity and surpassing genius. Prior to the discovery of those laws the Sun, though acknowledged to be the centre of the system, did not appear to occupy a central position as regards the motions of the planets; but Kepler, by demonstrating that the planes of the orbits of all the planets, and the lines connecting their apsides, passed through the Sun, was enabled to assign the orb his true position with regard to those bodies.

JOHN KEPLER was born at Weil, in the Duchy of Wurtemberg, December 21, 1571. His parents, though of noble family, lived in reduced circumstances, owing to causes for which they were themselves chiefly responsible. In his youth Kepler suffered so much from ill-health that his education had to be neglected. In 1586 he was sent to a monastic school at Maulbronn, which had been established at the Reformation, and was under the patronage of the Duke of Wurtemberg. Afterwards he studied at the University of Tubingen, where he distinguished himself and took a degree. Kepler devoted his attention chiefly to science and mathematics, but paid no particular attention to the study of astronomy. Maestlin, the professor of mathematics, whose lectures he attended, upheld the Copernican theory, and Kepler, who adopted the views of his teacher, wrote an essay in favour of the diurnal rotation of the Earth, in which he supported the more recent astronomical doctrines. In 1594, a vacancy having occurred in the professorship of astronomy at Gratz consequent upon the death of George Stadt, Kepler was appointed his successor. He did not seek this office, as he felt no particular desire to take up the study of astronomy, but was recommended by his tutors as a man well fitted for the post. He was thus in a manner compelled to devote his time and talents to the science of astronomy. Kepler directed his attention to three subjects—viz. 'the number, the size, and the motion of the orbits of the planets.' He endeavoured to ascertain if any regular proportion existed between the sizes of the planetary orbits, or in the difference of their sizes, but in this he was unsuccessful. He then thought that, by imagining the existence of a planet between Mars and Jupiter, and another between Venus and Mercury, he might be able to attain his object; but he found that this assumption afforded him no assistance.   Kepler then imagined that as there were five regular geometrical solids, and five planets, the distances of the latter were regulated by the size of the solids described round one another. The discovery afterwards of two additional planets testified to the absurdity of this speculation.   A description of these extraordinary researches was published, in 1596, in a work entitled ' Prodromus of Cosmographical Dissertations ; containing the cosmographical mystery respecting the admirable proportion of the celestial orbits, and the genuine and real causes of the number, magnitude, and periods of the planets, demonstrated by the five regular geometrical solids.'   This volume, notwith-standing the fanciful speculations which it contained, was received with much favour by astronomers, and both Tycho Brahé and Galileo encouraged Kepler to continue his researches.   Galileo admired his ingenuity, and Tycho advised him 'to lay a solid foundation for his views by actual observation, and then, by ascending from these, to strive to reach the causes of things.'   Kepler spent many years in these fruitless endeavours before he made those grand discoveries in search of which he laboured so long.

The religious dissensions which at this time agitated Germany were accompanied in many places by much tumult and excitement.   At Gratz the Catholics threatened to expel the Protestants from the city. Kepler, who was of the Reformed faith, having recognised the danger with which he was threatened, retired to Hungary with his wife, whom he had recently married, and remained there for near twelve months, during which time he occupied himself with writing several short treatises on subjects connected with astronomy. In 1599 he returned to Gratz and resumed his professorship.

In the year 1600 Kepler set out to pay Tycho Brahé a visit at Prague, in order that he might be able to avail himself of information contained in observations made by Tycho with regard to the eccentricities of the orbits of the planets. He was received by Tycho with much cordiality, and stayed with him for four months at his residence at Benach, Tycho in the meantime having promised that he would use his influence with the Emperor Rudolph to have him appointed as assistant in his observatory. On the termination of his visit Kepler returned to Gratz, and as there was a renewal of the religious trouble in the city, he resigned his professorship, from which he only derived a small income, and, relying on Tycho's promise, he again journeyed to Prague, and arrived there in 1601. Kepler was presented to the Emperor by Tycho, and the post of Imperial Mathematician was conferred upon him, with a salary of 100 florins a year, upon condition that he should assist Tycho in his observatory. This appointment was of much value to

Kepler, because it afforded him an opportunity of obtaining access to the numerous astronomical observations made by Tycho, which were of great assistance to him in the investigation of the subject which he had chosen—viz. the laws which govern the motions of the planets, and the form and size of the planetary orbits.

As an acknowledgment of the Emperor's great kindness, the two astronomers resolved to compute a new set of astronomical tables, and in honour of his Majesty they were to be called the 'Rudolphine Tables.' This project pleased the Emperor, who promised to defray the expense of their publication. Logomontanus, Tycho's chief assistant, had entrusted to him that portion of the work relating to observations on the stars, and Kepler had charge of the part which embraced the calculations belonging to the planets and their orbits. This important work had scarcely been begun when the departure of Logomontanus, who obtained an appointment in Denmark, and the death of Tycho Brahé in October 1601, necessitated its suspension for a time. Kepler was appointed Chief Mathematician to the Emperor in succession to Tycho—a position of honour and distinction, and to which was attached a handsome salary, that was paid out of the Imperial treasury. But owing to the continuance of expensive wars, which entailed a severe drain upon the resources of the country, the public funds became very low, and Kepler's salary was always in arrear. This condition of things involved him in serious pecuniary difficulties, and the responsibility of having to maintain an increasing family added to his anxieties. It was with the greatest difficulty that he succeeded in obtaining payment of even a portion of his salary, and he was reduced to such straits as to be under the necessity of casting nativities in order to obtain money to meet his most pressing requirements.

In 1609 Kepler published his great work, entitled 'The New Astronomy; or, Commentaries on the Motions of Mars.' It was by his observation of Mars, which has an orbit of greater eccentricity than that of any of the other planets, with the exception of Mercury, that he was enabled, after years of patient study, to announce in this volume the discovery of two of the three famous theorems known as Kepler's Laws. The first is, that all the planets move round the Sun in elliptic orbits, and that the orb occupies one of the foci. The second is, that the radius-vector, or imaginary line joining the centre of the planet and the centre of the Sun, describes equal areas in equal times. The third law, which relates to the connection between the periodic times and the distances of the planets, was not discovered until ten years later, when Kepler, in 1619, issued another work, called the 'Harmonies of the World,' dedicated to James I. of England, in which was contained this remarkable law. These laws have elevated astronomy to the position of a true physical science, and also formed the starting-point of Newton's investigations which led to the discovery of

the law of gravitation. Kepler's delight on the
discovery of his third law was unbounded. He
writes : ' Nothing holds me. I will indulge in my
sacred fury. I will triumph over mankind by the
honest confession that I have stolen the golden
vases of the Egyptians to build up a tabernacle for
my God far away from the confines of Egypt. If
you forgive me, I rejoice ; if you are angry, I can
bear it. The die is cast ; the book is written, to be
read either now or by posterity I care not which.
It may well wait a century for a reader, as God has
waited six thousand years for an observer.'

When Kepler presented his celebrated book to
the Emperor, he remarked that it was his intention
to make a similar attack upon the other planets,
and promised that he would be successful if his
Majesty would undertake to find the means necessary
for carrying on operations. But the Emperor had
more formidable enemies to contend with nearer
home than Jupiter and Saturn, and no funds were
forthcoming to assist Kepler in his undertaking.

The chair of mathematics in the University of
Linz having become vacant, Kepler offered himself
as a candidate for the appointment, which he was
anxious to obtain ; but the Emperor Rudolph was
averse to his leaving Prague, and encouraged him
to hope that the arrears of his salary would be paid.
But past experience led Kepler to have no very sanguine expectations on this point ; nor was it until
after the death of Rudolph, in 1612, that he was
relieved from his pecuniary embarrassments.

On the accession of Rudolph's brother, Matthias, to the Austrian throne, Kepler was reappointed Imperial Mathematician ; he was also permitted to hold the professorship at Linz, to which he had been elected. Kepler was not loth to remove from Prague, where he had spent eleven years harassed by poverty and other domestic afflictions. Having settled with his family at Linz, Kepler issued another work, in 1618, entitled 'Epitome of the Copernican Astronomy,' in which he gave a general account of his astronomical observations and discoveries, and a summary of his opinions with regard to the theories which in those days were the subject of controversial discussion. Almost immediately after its publication it was included by the Congregation of the Index, at Rome, in the list of prohibited books. This occasioned Kepler considerable alarm, as he imagined it might interfere with the sale of his works, or give rise to difficulties in the issue of others. He, however, was assured by his friend Remus that the action of the Papal authorities need cause him no anxiety.

The Emperor Matthias died in 1619, and was succeeded by Ferdinand III., who not only retained Kepler in his office, but gave orders that all the arrears of his salary should be paid, including those which accumulated during the reign of Rudolph ; he also expressed a desire that the 'Rudolphine Tables' should be published without delay and at his cost. But other obstacles intervened, for at this time Germany was involved in a civil and religious war, which interfered with all peaceful vocations. Kepler's library at Linz was sealed up by order of the Jesuits, and the city was for a time besieged by troops.   This state of public affairs necessitated a considerable delay in the publication of the ' Tables.'

The ' Rudolphine Tables ' were published at Ulm in 1627.   They were commenced by Tycho Brahé, and completed by Kepler, who made his calculations from Tycho's observations, and based them upon his own great discovery of the ellipticity of the orbits of the planets.   They are divided into four parts.   The first and third parts contain logarithmic and other tables for the purpose of facilitating astronomical calculations; in the second are tables of the Sun, Moon, and planets; and in the fourth are indicated the positions of one thousand stars as determined by Tycho.   Kepler made a special journey to Prague in order to present the ' Tables ' to the Emperor, and afterwards the Grand Duke of Tuscany sent him a gold chain as an acknowledgment of his appreciation of the completion of this great work.

Albert Wallenstein, Duke of Friedland, an accomplished scholar and a man fond of scientific pursuits, made Kepler a most liberal offer if he would take up his residence in his dominions. After duly considering this proposal, Kepler decided to accept the Duke's offer, provided it received the sanction of the Emperor.   This was readily given, and Kepler, in 1629, removed with his family from Linz to Sagan, in Silesia.   The Duke of Friedland treated him with great kindness and liberality, and through his influence he was appointed to a professorship in the University of Rostock. Though Kepler was permitted to retain the pension bestowed upon him by the late Emperor Rudolph, he was unable after his removal to Silesia to obtain payment of it, and there was a large accumulation of arrears. In a final endeavour to recover the amount owing to him he travelled to Ratisbon, and appealed to the Imperial Assembly, but without success. The fatigue which Kepler endured on his journey, combined with vexation and disappointment, brought on a fever, which terminated fatally. He died on November 15, 1630, when in the sixtieth year of his age, and was interred in St. Peter's churchyard, Ratisbon.

Kepler was a man of indomitable energy and perseverance, and spared neither time nor trouble in the accomplishment of any object which he took in hand. In thinking over the form of the orbits of the planets, he writes : ' I brooded with the whole energy of my mind on this subject—asking why they are not other than they are—the number, the size, and the motions of the orbits.' But many fanciful ideas passed through Kepler's imaginative brain before he hit upon the true form of the planetary orbits. In his ' Mysterium Cosmographicum ' he asserts that the five kinds of regular polyhedral solids, when described round one another, regulated the distances of the planets and size of the planetary orbits. In support of this theory he writes as follows : 'The orbit of the Earth is the measure of the rest.   About it circumscribe a dode-cahedron.   The sphere including this will be that of Mars.   About Mars' orbit describe a tetrahedron ; the sphere containing this will be Jupiter's orbit. Round Jupiter's describe a cube ; the sphere in-cluding this will be Saturn's.   Within the Earth's orbit inscribe an icosahedron ; the sphere inscribed in it will be Venus's orbit.   In Venus inscribe an octahedron ; the sphere inscribed in it will be Mercury's.'

The above quotation is an instance of Kepler's wild and imaginative genius, which ultimately led him to make those sublime discoveries associated with planetary motion which are known as 'Kepler's Laws.'

He describes himself as 'troublesome and choleric in politics and domestic matters ; ' but in his relations with scientific men he was affable and pleasant.   He showed no jealousy of a rival, and was always ready to recognise merit in others ; nor did he hesitate to acknowledge any error of his own when more recent discoveries proved that he was wrong.

Some of his works contain passages, written in a jocular strain, indicative of a bright and cheerful temperament.   The following characteristic para-graph refers to the opinions of the Epicureans with regard to the appearance of a new star, which they ascribed to a fortuitous concourse of atoms : 'When I was a youth, with plenty of idle time on my hands, I was much taken with the vanity, of which some grown men are not ashamed, of making anagrams by transposing the letters of my name written in Latin so as to make another sentence. Out of Ioannes Keplerus came *Serpens in akuleo* (a serpent in his sting) ; but not being satisfied with the meaning of these words, and being unable to make another, I trusted the thing to chance, and, taking out of a pack of playing-cards as many as there were letters in the name, I wrote one upon each, and then began to shuffle them, and at each shuffle to read them in the order they came, to see if any meaning came of it. Now, may all the Epicurean gods and goddesses confound this same chance, which, although I have spent a good deal of time over it, never showed me anything like sense, even from a distance. So I gave up my cards to the Epicurean eternity, to be carried away into infinity ; and it is said they are still flying about there, in the utmost confusion, among the atoms, and have never yet come to any meaning. I will tell those disputants, my opponents, not my own opinion, but my wife's. Yesterday, when weary with writing, and my mind quite dusty with considering these atoms, I was called to supper, and a salad I had asked for was set before me. " It seems, then," said I aloud, " that if pewter dishes, leaves of lettuce, grains of salt, drops of water, vinegar and oil, and slices of egg, had been flying about in the air from all eternity, it might at last happen by chance that there would come a salad."

D

" Yes," says my wife, " but not so nice and well dressed as this of mine is." '

Notwithstanding the frequent interruptions which, owing to various reasons, retarded his labours, Kepler was able to bring to a successful completion the numerous and important works upon which he was engaged during his lifetime, the voluminous nature of which may be imagined when it is stated that he published thirty-three separate works, besides leaving behind twenty-two volumes of manuscript.

During his researches on the motions of Mars, Kepler discovered that the planet sometimes travelled at an accelerated rate of speed, and at another time its pace was diminished. At one time he observed it to be in advance of the place where he calculated it should be found, and at another time it was behind it. This caused him considerable perplexity, and, feeling convinced in his mind that the form of the planet's orbit could not be circular, he was compelled to turn his attention to some other closed curve, by which those inequalities of motion could be explained.

After years of careful observation and study, Kepler arrived 'at the conclusion that the form of the planet's orbit is an ellipse, and that the Sun occupies one of the foci. He afterwards determined that the orbits of all the planets are of an elliptical form.

Having discovered the true form of the planetary orbits, Kepler next endeavoured to ascertain the cause which regulates the unequal motion that a planet pursues in its path. He observed that when a planet approached the Sun its motion was accelerated, and as it receded from him its pace became slower.

This he explained in his next great discovery by proving that an imaginary line, or radius-vector, extending from the centre of the Sun to the centre of the planet 'describes equal areas in equal times.' When near the Sun, or at perihelion, a planet traverses a larger portion of its arc in the same period of time than it does when at the opposite part of its orbit, or when at aphelion; but, as the areas of both are equal, it follows that the planet does not always maintain the same rate of speed, and that its velocity is greatest when nearest the Sun, and least when most distant from him.

By the application of his first and second laws Kepler was able to formulate a third law. He found that there existed a remarkable relationship between the mean distances of the planets and the times in which they complete their revolutions round the Sun, and discovered 'that the squares of the periodic times are to each in the same proportion as the cubes of the mean distances.' The periodic time of a planet having been ascertained, the square of the mean distance and the mean distance itself can be obtained. It is by the application of this law that the distances of the planets are usually calculated.

These discoveries are known as Kepler's Laws, and are usually classified as follows :—

1. 'The orbit described by every planet is an ellipse, of which the centre of the Sun occupies one of the foci.

2. 'Every planet moves round the Sun in a plane orbit, and the radius-vector, or imaginary line joining the centre of the planet and the centre of the Sun, describes equal areas in equal times.

3. 'The squares of the periodic times of any two planets are proportional to the cubes of their mean distances from the Sun.' [1]

These remarkable discoveries do not embrace all the achievements by which Kepler has immortalised his name, and earned for himself the proud title of 'Legislator of the Heavens;' he predicted transits of Mercury and Venus, made important discoveries in optics, and was the inventor of the astronomical telescope.

GALILEO GALILEI, the famous Italian astronomer and philosopher, and the contemporary of Kepler and of Milton, was born at Pisa on February 15, 1564.

His father, who traced his descent from an ancient Florentine family, was desirous that his son should adopt the profession of medicine, and with this intention he entered him as a student at the University of Pisa. Galileo, however, soon discovered that the study of mathematics and mechanical science possessed a greater attraction

---

[1] Chambers's *Handbook of Astronomy.*

for his mind, and, following his inclinations, he resolved to devote his energies to acquiring proficiency in those subjects.

In 1583 his attention was attracted by the oscillation of a brass lamp suspended from the ceiling of the cathedral at Pisa. Galileo was impressed with the regularity of its motion as it swung backwards and forwards, and was led to imagine that the pendulum movement might prove a valuable method for the correct measurement of time. The practical application of this idea he afterwards adopted in the construction of an astronomical clock.

Having become proficient in mathematics, Galileo, whilst engaged in studying the writings of Archimedes, wrote an essay on ' The Hydrostatic Balance,' and composed a treatise on ' The Centre of Gravity in Solid Bodies.' The reputation which he earned by these contributions to science procured for him the appointment of Lecturer on Mathematics at the University of Pisa. Galileo next directed his attention to the works of Aristotle, and made no attempt to conceal the disfavour with which he regarded many of the doctrines taught by the Greek philosopher ; nor had he any difficulty in exposing their inaccuracies. One of these, which maintained that the heavier of two bodies descended to the earth with the greater rapidity, he proved to be incorrect, and demonstrated by experiment from the top of the tower at Pisa that, except for the unequal resistance of the air, all bodies fell to the ground with the same velocity.

As the chief expounder of the new philosophy, Galileo had to encounter the prejudices of the followers of Aristotle, and of all those who disliked any innovation or change in the established order of things.   The antagonism which existed between Galileo and his opponents, who were both nume-rous and influential, was intensified by the bitter-ness and sarcasm which he imparted into his controversies, and the attitude assumed by his enemies at last became so threatening that he deemed it prudent to resign the Chair of Mathe-matics in the University of Pisa.

In the following year he was appointed to a similar post at Padua, where his fame attracted crowds of pupils from all parts of Europe.

In 1611 Galileo visited Rome.   He was received with much distinction by the different learned societies, and was enrolled a member of the Lyncæan Academy.   In two years after his visit to the capital he published a work in which he declared his adhesion to the Copernican theory, and openly avowed his disbelief in the astronomical facts recorded in the Scriptures.   Galileo main-tained that the sacred writings were not intended for the purpose of imparting scientific informa-tion, and that it was impossible for men to ignore phenomena witnessed with their eyes, or disre-gard conclusions arrived at by the exercise of their reasoning powers.

The champions of orthodoxy having become alarmed, an appeal was made to the ecclesiastical authorities to assist in suppressing this recent astronomical heresy, and other obnoxious doctrines, the authorship of which was ascribed to Galileo.

In 1615, Galileo was summoned before the Inquisition to reply to the accusation of heresy. ' He was charged with maintaining the motion of the Earth and the stability of the Sun; with teaching this doctrine to his pupils ; with corresponding on the subject with several German mathematicians ; and with having published it, and attempted to reconcile it to Scripture in his letters to Mark Velser in 1612.'

These charges having been formally investigated by the Inquisition, Cardinal Bellarmine was authorised to communicate with Galileo, and inform him that unless he renounced the obnoxious doctrines, and promised ' neither to teach, defend, or publish them in future,' it was decreed that he should be committed to prison. Galileo appeared next day before the Cardinal, and, without any hesitation, pledged himself that for the future he would adhere to the pronouncement of the Inquisition.

Having, as they imagined, silenced Galileo, the Inquisition resolved to condemn the entire Copernican system as heretical ; and in order to effectually accomplish this, besides condemning the writings of Galileo, they inhibited Kepler's ' Epitome of the Copernican System,' and Copernicus's own work, ' De Revolutionibus Orbium Celestium.'

Whether it was that Galileo regarded the Inquisition as a body whose decrees were too

absurd and unreasonable to be heeded, or that he
dreaded the consequences which might have followed had he remained obstinate, we know that,
notwithstanding the pledges which he gave, he was
soon afterwards engaged in controversial discussion
on those subjects which he promised not to mention
again.

On the accession of his friend Cardinal Barberini
to the pontifical throne in 1623, under the title of
Urban VIII., Galileo undertook a journey to Rome
to offer him his congratulations upon his elevation
to the papal chair. He was received by his Holi-
ness with marked attention and kindness, was
granted several prolonged audiences, and had conferred upon him several valuable gifts.

Notwithstanding the kindness of Pope Urban
and the leniency with which he was treated by the
Inquisition, Galileo, having ignored his pledge,
published in 1632 a book, in dialogue form, in
which three persons were supposed to express their
scientific opinions. The first upheld the Copernican
theory and the more recent philosophical views;
the second person adopted a neutral position, suggested doubts, and made remarks of an amusing
nature ; the third individual, called Simplicio, was
a believer in Ptolemy and Aristotle, and based his
arguments upon the philosophy of the ancients.

As soon as this work became publicly known,
the enemies of Galileo persuaded the Pope that
the third person held up to ridicule was intended
as a representation of himself—an individual

regardless of scientific truth, and firmly attached to the ideas and opinions associated with the writings of antiquity.

Almost immediately after the publication of the ' Dialogues ' Galileo was summoned before the Inquisition, and, notwithstanding his feeble health and the infirmities of advanced age, he was, after a long and tedious trial, condemned to abjure by oath on his knees his scientific beliefs.

' The ceremony of Galileo's abjuration was one of exciting interest and of awful formality. Clothed in the sackcloth of a repentant criminal, the venerable sage fell upon his knees before the assembled cardinals, and, laying his hand upon the Holy Evangelists, he invoked the Divine aid in abjuring, and detesting, and vowing never again to teach the doctrines of the Earth's motion and of the Sun's stability. He pledged himself that he would nevermore, either in words or in writing, propagate such heresies ; and he swore that he would fulfil and observe the penances which had been inflicted upon him.' ' At the conclusion of this ceremony, in which he recited his abjuration word for word and then signed it, he was conveyed, in conformity with his sentence, to the prison of the Inquisition.' [1]

Galileo's sarcasm, and the bitterness which he imparted into his controversies, were more the cause of his misfortunes than his scientific beliefs. When he became involved in difficulties he did not possess the moral courage to enable him to abide

[1] Brewster's *Martyrs of Science.*

by the consequences of his acts ; nor did he care to
become a martyr for the sake of science, his sub-
mission to the Inquisition having probably saved
him from a fate similar to what befell Bruno.
Though it would be impossible to justify Galileo's
want of faith in his dealings with the Inquisition,
yet one cannot help sympathising deeply with the
aged philosopher, who, in this painful episode of
his life, was compelled to go through the form of
making a retractation of his beliefs under circumstances of a most humiliating nature.

But the persecution of Galileo did not delay
the progress of scientific inquiry nor retard the
advancement of the Copernican theory, which,
after the discovery by Newton of the law of gravitation, was universally adopted as the true theory
of the solar system.

Ferdinand, Duke of Tuscany, having exerted
his influence with Pope Urban on behalf of Galileo,
he was, after a few days' incarceration, released
from prison, and permission was given him to
reside at Siena, where he remained for six months.
He was afterwards allowed to return to his villa at
Arcetri, and, though regarded as a prisoner of the
Inquisition, was permitted to pursue his studies
unmolested for the remainder of his days.

Galileo died at Arcetri on January 8, 1642,
when in the seventy-eighth year of his age.

Though not the inventor, he was the first to
construct a refracting telescope and apply it to
astronomical research. With this instrument he

made a number of important discoveries which tended to confirm his belief in the truthfulness of the Copernican theory.

On directing his telescope to the Sun, he discovered movable spots on his disc, and concluded from his observation of them that the orb rotated on his axis in about twenty-eight days. He also ascertained that the Moon's illumination is due to reflected sunlight, and that her surface is diversified by mountains, valleys, and plains.

On the night of January 7, 1610, Galileo discovered the four moons of Jupiter. This discovery may be regarded as one of his most brilliant achievements with the telescope; and, notwithstanding the improvement in construction and size of modern instruments, no other satellite was discovered until near midnight on September 9, 1892, when Mr. E. E. Barnard, with the splendid telescope of the Lick Observatory, added 'another gem to the diadem of Jupiter.'

The phases of Venus and Mars, the triple form of Saturn, and the constitution of the Milky Way, which he found to consist of a countless multitude of stars, were additional discoveries for our knowledge of which we are indebted to Galileo and his telescope. Galileo made many other important discoveries in mechanical and physical science. He detected the law of falling bodies in their accelerated motion towards the Earth, determined the parabolic law of projectiles, and demonstrated that matter, even if invisible, possessed the property of weight.

In these pages a short historical description is given of the progress made in astronomical science from an early period to the time in which Milton lived. The discoveries of Copernicus, Kepler, and Galileo had raised it to a position of lofty eminence, though the law of gravitation, which accounts for the form and permanency of the planetary orbits, still remained undiscovered. Theories formerly obscure or conjectural were either rejected or elucidated with accuracy and precision, and the solar system, having the Sun as its centre, with his attendant family of planets and their satellites revolving in majestic orbits around him, presented an impressive spectacle of order, harmony, and design.

CHAPTER II

THE seventeenth century embraces the most remarkable epoch in the whole history of astronomy. It was during this period that those wonderful discoveries were made which have been the means of raising astronomy to the lofty position which it now occupies among the sciences. The unrivalled genius and patient labours of the illustrious men whose names stand out in such prominence on the written pages of the history of this era have rendered it one of the most interesting and elevating of studies. Though Copernicus lived in the preceding century, yet the names of Tycho Brahé, Kepler, Galileo, and Newton, testify to the greatness of the discoveries that were made during this period, which have surrounded the memories of those men with a lustre of undying fame.

Foremost among astronomers of less conspicuous eminence who made important discoveries in this century we find the name of Huygens.

CHRISTIAN HUYGENS was born at The Hague in 1629. He was the second son of Constantine Huygens, an eminent diplomatist, and secretary to the Prince of Orange. Huygens studied at Leyden

and Breda, and became highly distinguished as a
geometrician and scientist. He made important
investigations relative to the figure of the Earth,
and wrote a learned treatise on the cause of gravity ;
he also determined with greater accuracy investigations made by Galileo regarding the accelerated
motion of bodies when subjected to the influence of
that force.

Huygens admitted that the planets and their
satellites attracted each other with a force varying
according to the inverse ratio of the squares of their
distances, but rejected the ı ıutua¹ attraction of the
molecules of matter, believing that they possessed
gravity towards a central point only, to which they
were attracted. This supposition was at variance
with the Newtonian theory, which, however, was
universally regarded as the correct one.

Huygens originated the theory by which it is
believed that light is produced by the undulatory
vibration of the ether ; he also discovered polariza-
tion.

Up to this time the method adopted in the construction of clocks was not capable of producing a
mechanism which measured time with sufficient
accuracy to satisfy the requirements of astronomers.
Huygens endeavoured to supply this want, and
applied his mechanical ingenuity in constructing a
clock that could be relied upon to keep accurate
time. Though the pendulum motion was first
adopted by Galileo, he was unable to arrange its
mechanism so that it should keep up a continuous
movement. The oscillation of the pendulum ceased

after a time, and a fresh impulse had to be applied to set it in motion. Consequently, Galileo's clock was of no service as a timekeeper.

Huygens overcame this difficulty by so arranging the mechanism of his clock that the balance, instead of being horizontal, was directed perpendicularly, and prolonged downwards to form a pendulum, the oscillations of which regulated the downward motion of the weight. This invention, which was highly applauded, proved to be of great service everywhere, and was especially valuable for astronomical purposes.

Huygens next directed his attention to the construction of telescopes, and displayed much skill in the grinding and polishing of lenses. He made several instruments superior in power and accuracy to any that existed previously, and with one of these made some remarkable discoveries when observing the planet Saturn.

The telescopic appearance of Saturn is one of the most beautiful in the heavens. The planet, surrounded by two brilliant rings, and accompanied by eight attendant moons, surpasses all the other orbs of the firmament as an object of interest and admiration. To the naked eye, Saturn is visible as a star of the first magnitude, and was known to the ancients as the most remote of the planets. Travelling in space at a distance of nearly one thousand millions of miles from the Sun, the planet accomplishes a revolution of its mighty orbit in twenty-nine and a half years.

Galileo was the first astronomer who directed a

telescope to Saturn. He observed that the planet presented a triform appearance, and that on each side of the central globe there were two objects, in close contact with it, which caused it to assume an ovoid shape. After further observation, Galileo perceived that the lateral bodies gradually decreased in size, until they became invisible. At the expiration of a certain period of time they reappeared, and were observed to go through a certain cycle of changes. By the application of increased telescopic power it was discovered that the appendages were not of a rounded form, but appeared as two small crescents, having their concave surfaces directed towards the planet and their extremities in contact with it, resembling the manner in which the handles are attached to a cup.

These objects were observed to go through a series of periodic changes. After having become invisible, they reappeared as two luminous straight bands, projecting from each side of the planet ; during the next seven or eight years they gradually opened out, and assumed a crescentic form ; they afterwards began to contract, and on the expiration of a similar period, during which time they gradually decreased in size, they again became invisible. It was perceived that the appendages completed a cycle of their changes in about fifteen years.

In 1656, Huygens, with a telescope constructed by himself, was enabled to solve the enigma which for so many years baffled the efforts of the ablest astronomers. He announced his discovery in the form of a Latin cryptograph which, when deciphered, read as follows :—

'Annulo cingitur, tenui plano, nusquam cohaerente, ad eclipticam inclinatio.'

'The planet is surrounded by a slender flat ring everywhere distinct from its surface, and inclined to the ecliptic.'

Huygens perceived the shadow of the ring thrown on the planet, and was able to account in a satisfactory manner for all the phenomena observed in connection with its variable appearance.

The true form of the ring is circular, but by us it is seen foreshortened; consequently, when the Earth is above or below its plane, it appears of an elliptical shape. When the position of the planet is such that the plane of the ring passes through the Sun, the edge of the ring only is illumined, and then it becomes invisible for a short period. In the same manner, when the plane of the ring passes through the Earth, the illumined edge of the ring is not of sufficient magnitude to appear visible, but as the enlightened side of the plane becomes more inclined towards the Earth, the ring comes again into view. When the plane of the ring passes between the Earth and the Sun, the unillumined side of the ring is turned towards the Earth, and during the time it remains in this position it is invisible.

Huygens discovered the sixth satellite of Saturn (Titan), and also the Great Nebula in Orion.

JOHANN HEVELIUS, a celebrated Prussian astronomer, was born at Dantzig in 1611, and died in

E

that city in 1687. He was a man of wealth, and erected an observatory at his residence, where, for a period of forty years, he carried out a series of astronomical observations.

He constructed a chart of the stars, and in order to complete his work, formed nine new constellations in those spaces in the celestial vault which were previously un-named. They are known by the names Camelopardus, Canes Venatici, Coma Bernices, Lacerta, Leo Minor, Lynx, Monoceros, Sextans, and Vulpecula. He also executed a chart of the Moon's surface, wrote a description of the lunar spots, and discovered the Libration of the Moon in Longitude.

On May 30, 1661, Hevelius observed a transit of Mercury, a description of which he published, and included with it Horrox's treatise on the first-recorded transit of Venus. This work, after having passed through several hands, became the property of Hevelius, who was capable of appreciating its merits. The manuscript was sent to him by Huygens, and in acknowledging it he writes : ' How greatly does my Mercury exult in the joyous prospect that he may shortly fold within his arms Horrox's long looked-for and beloved Venus! He renders you unfeigned thanks that by your permission this much-desired union is about to be celebrated, and that the writer is able, with your concurrence, to introduce them both together to the public.'

Hevelius made numerous researches on comets,

and suggested that the form of their paths might be a parabola.

GIOVANNI DOMENICO CASSINI was born at Perinaldo, near Nice, in 1625. He studied at Genoa and Bologna, and was afterwards appointed to the Chair of Astronomy at the latter University. He was a man of high scientific attainments, and made many important astronomical discoveries.

In 1671 he became Director of the Royal Observatory at Paris, and devoted a long life to trying and difficult observations, which in his later years deprived him of his eyesight.

In 1644 Cassini proved beyond doubt that Jupiter rotated on his axis, and also assigned his period of rotation with considerable accuracy. He published tables of the planet's satellites, and determined their motions from observations of their eclipses. He ascertained the periods of rotation of Venus and Mars ; executed a chart of the lunar surface, and observed an occultation of Jupiter by the Moon.

Cassini discovered the dual nature of Saturn's ring, having perceived that instead of one there are two concentric rings separated by a dark space. He also discovered four of the planet's satellites— viz. Japetus, Rhea, Dione, and Tethys. He made a near approximation to the solar parallax by means of researches on the parallax of Mars, and investigated some irregularities of the Moon's motion. Cassini discovered the belts of Jupiter, and also the

Zodiacal Light, and established the coincidence of the nodes of the lunar equator and orbit.

JAQUES CASSINI, son of Giovanni, was born at Paris in 1677. He followed in his father's footsteps, and wrote several treatises on astronomical subjects. He investigated the period of the rotation of Venus on her axis, and upheld the results arrived at by his father, which were afterwards confirmed by observations made by Schroeter. Cassini made some valuable researches with regard to the proper motion of the stars, and demonstrated that their change of position on the celestial vault was real, and not caused by a displacement of the ecliptic. He attempted to ascertain the apparent diameter of Sirius, and made observations with regard to the visibility of the stars. The Cassini family produced several generations of eminent astronomers, whose discoveries and investigations were of much value in advancing the science of astronomy.

OLAUS ROEMER, an eminent Danish astronomer, was born at Copenhagen September 25, 1644. When Picard, a French astronomer, visited Denmark in 1671, for the purpose of ascertaining the exact position of 'Uranienburg,' the site of Tycho Brahé's observatory, he made the acquaintance of Roemer, who was engaged in studying mathematics and astronomy under Erasmus Bartolinus. Having perceived that the young man was gifted with no ordinary degree of talent, he secured his services to assist him in his observations, and, on the conclusion of his labours, Picard was so much impressed with the ability displayed by Roemer, that he invited him to accompany him to France. This invitation he accepted, and took up his residence in the French capital, where he continued to prosecute his astronomical studies.

In 1675 Roemer communicated to the Academy of Sciences a paper, in which he announced his discovery of the progressive transmission of light. It was believed that light travelled instantaneously, but Roemer was able to demonstrate the inaccuracy of this conclusion, and determined that light travels through space with a measurable velocity.

By diligently observing the eclipses of Jupiter's satellites, Roemer perceived that sometimes they occurred before, and sometimes after their predicted times. This irregularity, he discovered, depended upon the position of the Earth with regard to Jupiter. When the Earth, in traversing her orbit, moved round to the opposite side of the Sun, thereby bringing Jupiter into conjunction, an eclipse occurred sixteen minutes twenty-six seconds later than it did when Jupiter was in opposition or nearest to the Earth. As there existed an impression that light travelled instantaneously, it was believed that an eclipse occurred at the moment it was perceived in the telescope. This, however, was not so. Roemer, after a long series of observations, concluded that the discrepancies were due to the fact that light travels with a measurable velocity, and that it requires a greater length of time, upwards of sixteen minutes, to traverse the additional distance—the diameter of the Earth's orbit—which intervenes between the Earth and Jupiter, when the planet is in conjunction, as compared with the distance between the Earth and Jupiter, when the latter is in opposition. This discovery of Roemer's was the means of enabling the velocity of light to be ascertained, which, according to recent calculations, is about 187,000 miles a second. As an acknowledgment of the importance of his communication, Roemer was awarded a seat in the Academy, and apartments were assigned to him at the Royal Observatory, where he carried on his astronomical studies.

In 1681 Roemer returned to Denmark, and was appointed Professor of Mathematics in the University of Copenhagen ; he was also entrusted with the care of the city observatory—a duty which his reputation as an astronomer eminently qualified him to undertake. The transit instrument—a mechanism of much importance to astronomers—was invented by Roemer in 1690 ; it consists of a telescope fixed to a horizontal axis, and adjusted so as to revolve in the plane of the meridian. It is employed in observing the passage of the heavenly bodies across the observer's meridian. To note accurately by means of the astronomical clock the exact instant of time at which a celestial body crosses the centre of the field of view is the essential part of a transit observation. Small transit instruments are employed for taking the time and for regulating the observatory clock, but large instruments are used for delicate and exact observations of Right Ascensions and Declinations of stars of different magnitudes. Meridian, and altitude and azimuth circles, are important astronomical appliances, which owe their existence to the inventive skill of this distinguished astronomer.

Roemer resided for many years at the observatory in the city of Copenhagen, where he pursued his astronomical studies until the time of his death, which occurred in 1710. He meritoriously attempted to determine the parallax of the fixed stars; and it is said that the astronomical calculations and observations which he left behind him were so voluminous as to equal in number those made by Tycho Brahé, nearly all of which perished in a great conflagration that destroyed the observatory and a large portion of the city of Copenhagen in 1728.

Among other astronomers of this century whose names deserve recording were Descartes and Gassendi, whose mathematical researches in their application to astronomy were of much value; Fabricius, Torricelli, and Maraldi, who by their observations and investigations added many facts to the general knowledge of the science; and Bayer, to whom belongs the distinction of having constructed the first star-atlas.

In our own country during this period astronomy was cultivated by a few enthusiastic men, who devoted their time and talents to promoting the advancement of the science. It, however, received no recognition as a subject of study at any of the Universities, and no public observatory existed in Great Britain.

Though it was not until towards the close of the century that the attention of all Europe was directed to England in admiration of the discoveries of the illustrious Newton, yet astronomy had its humble votaries, and chief among those was a young clergyman of the name of Horrox.

JEREMIAH HORROX was born at Toxteth, near Liverpool, in 1619—close on three centuries ago. Little is known of his family. His parents have been described as persons who occupied a humble position in life, but, as they were able to give their son a classical education which fitted him for one of the learned professions, it is probable they were not so obscure as they have been represented to be.

Having received his early education at Toxteth, Horrox afterwards proceeded to Cambridge, and was entered as a student at Emmanuel College on May 18, 1632, when in his fourteenth year.

At the University he devoted himself to the study of classics, especially Latin, which in those days was the language adopted by men of learning, when engaged in writing works of a philosophical and scientific character.

After having remained at Cambridge for three years, Horrox returned to his native county, and was appointed curate of Hoole, a place about eight miles distant from Preston. Hoole is described as a narrow low-lying strip of land consisting largely of moss, and almost converted into an island by the waters of Martin Mere on the south, and the Ribble on the north ; and, though doubtless an open and favourable situation for astronomical observation, it could not have been attractive as a place of residence. Yet it was here on November 24, 1639, that Horrox made his famous observation of the first recorded transit of Venus, an occurrence with which his name will be for ever associated.

It was while at Cambridge that Horrox first turned his attention to the study of astronomy. His love of the sublime, and the captivating influence exerted on his mind by the contemplation of the heavenly bodies, induced him to adopt astronomy as a pursuit congenial to his tastes, and capable of exercising his highest mental powers. Having this object in view, he applied himself with much earnestness to the study of mathematics ; he had, however, to rely mainly upon his own exertions, for at that time no branch of physical or mathematical science was taught at Cambridge, and consequently he obtained no professional instruction.

It was so also with astronomy, which, as a science, was scarcely known in this country ; no regular record of astronomical observations was kept by any individual observer, and no public observatory existed in England or in France.

The disadvantages and obstacles which Horrox

had to encounter may be best described by quoting
his own words. He writes: 'There were many
hindrances. The abstruse nature of the study, my
inexperience and want of means dispirited me. I
was much pained not to have any one to whom I
could look for guidance, or indeed for the sympathy
of companionship in my endeavours, and I was
assailed by the languor and weariness which are
inseparable from every great undertaking. What
then was to be done ? I could not make the pursuit
an easy one, much less increase my fortune, and
least of all imbue others with a love for astronomy ;
and yet to complain of philosophy on account of
its difficulties would be foolish and unworthy. I
determined, therefore, that the tediousness of study
should be overcome by industry ; my poverty—
failing a better method—by patience ; and that
instead of a master I would use astronomical books.
Armed with these weapons I would contend successfully ; and, having heard of others acquiring
knowledge without greater help, I would blush that
any one should be able to do more than I, always
remembering that word of Virgil's—

> Totidem nobis animaeque manusque.

Having heard much praise bestowed upon the
works of Lansberg, a Flemish astronomer, Horrox
thought it would be to his advantage to procure a
copy of his writings. This he succeeded in obtaining after some difficulty, and devoted a considerable
time to calculating Ephemerides, based upon the

Lansberg Tables, but after making a number of computations he discovered that they were unreliable and inaccurate.

In the year 1636 Horrox made the acquaintance of William Crabtree, a devoted astronomer, who lived at Broughton, a suburb of Manchester. A close friendship soon existed between the two men, and they carried on an active correspondence about matters relating to the science which they both loved so well.

Crabtree, who was an unbeliever in Lansberg, urged Horrox to discard the Flemish astronomer's works, and devote his talents to the study of Tycho Brahé and Kepler. This advice led Horrox to make a more rigorous examination of the Lansberg Tables, and after comparing them with the observations made by Crabtree, which coincided with his own, he resolved to renounce them. Acting on the advice of his friend, Horrox directed his attention to the writings of Kepler. The youthful astronomer soon realised their value, and was charmed with the accuracy of observation and inductive reasoning displayed in the elucidation of those general laws which constituted a new era in the history of astronomy.

The Rudolphine Tables, which were the astronomical calculations commenced by Tycho Brahé, and completed by Kepler, were regarded by Horrox as much superior to those of Lansberg; but it occurred to him that they might be improved by changing some of the numbers, and yet retaining the hypotheses. To this task he applied himself with much earnestness and assiduity, and after close application and laborious study he accomplished the arduous undertaking of bringing those tables to a high state of perfection.

In his investigation of the Lunar theory, Horrox outstripped all his predecessors, and Sir Isaac Newton distinctly affirms he was the first to discover that the Moon's motion round the Earth is in the form of an ellipse with the centre in the lower focus. Besides having made this discovery, Horrox was able to explain the causes of the inequalities of the Moon's motion, which render the exact computation of her elements so difficult.

The Annual Equation, an irregularity discovered by Tycho Brahé, which is produced by the increase and decrease of the Sun's disturbing force as the Earth approaches or recedes from him in her orbit, had its value first assigned by Horrox. This he calculated to be eleven minutes sixteen seconds, which is within four seconds of what it has since been proved to be by the most recent observations.

The Evection, an irregular motion of the Moon discovered by Ptolemy, whereby her mean longitude is increased or diminished, was explained by Horrox as depending upon the libratory motion of the apsides, and the change which takes place in the eccentricity of the lunar orbit.

These discoveries were made by Horrox before he attained the age of twenty years, and if his reputation had alone rested upon them his name would have been honourably associated with those who have attained to the highest eminence in astronomy.

Another achievement which adds lustre to Horrox's name consists in his detection of the inequality in the mean motions of Jupiter and Saturn.

He also directed his attention to the study of cometary bodies, and arrived at certain conclusions with regard to the nature of their movements. At first, he believed like Kepler that comets were projected in straight lines from the Sun; this supposition having been upheld on account of the great elongation of their orbits. He next perceived that their velocity increased as they approached the Sun, and decreased as they receded from him. Afterwards he says, ' They move in an elliptic figure or near it,' and finally he arrived at the conclusion that ' comets move in elliptical orbits, being carried round the Sun with a velocity which is probably variable.' This theory has been verified by numerous observations, and is now generally accepted by astronomers.

Horrox also made a series of observations on the tides. He notified the extent of their rise and fall at different periods, and investigated other phenomena associated with their ebb and flow. After having continued his observations for some time, he wrote to his friend Crabtree, and informed him that he had perceived many interesting details which had not been previously described, and he

hoped to be able to arrive at some important conclusions with regard to their nature and cause.
Unfortunately, Horrox's writings on this subject,
along with many other important papers, have
been lost or destroyed. We are therefore ignorant
of the result of his researches, which were the first
undertaken by any person for the purpose of scientific inquiry.

From his study of the Lansberg and Rudolphine
Tables, Horrox arrived at the conclusion that a
transit of Venus would occur on November 24,
1639. This transit was for some unaccountable
reason overlooked by Kepler, who predicted one in
1631, and the next not until 1761. The transit of
1631 was not visible in Europe.

We are indebted to Horrox for a description of
the transit of 1639—the first that was ever observed
of which there is any record ; and were it not for the
accuracy of his calculations, the occurrence of the
phenomenon would have been unperceived, and no
history of the conjunction would have been handed
down to posterity. As soon as Horrox had assured
himself of the time when the transit would take
place, he wrote to Crabtree to inform him of the
date, and asked him to make observations with his
telescope, and especially to examine the diameter of
the planet, which he thought had been over-estimated. He also requested him to write to Dr.
Foster of Cambridge, and inform him of the expected
event, as it was desirable that the transit should be
observed from several places in consequence of the

possibility of failure, owing to an overcast sky.   His
letter is dated October 26, 1639.   He says : ' My
reason for now writing is to advise you of a remarkable conjunction of the Sun and Venus on
the 24th of November, when there will be a transit.
As such a thing has not happened for many years
past, and will not occur again in this century, I
earnestly entreat you to watch attentively with
your telescope in order to observe it as well as
you can.

' Notice particularly the diameter of Venus,
which is stated by Kepler to be seven minutes, and
by Lansberg to be eleven, but which I believe to
be scarcely greater than one minute.'

In describing the method which he adopted for
observing the transit, Horrox writes as follows :
' Having attentively examined Venus with my instrument, I described on a sheet of paper a circle,
whose diameter was nearly equal to six inches—
the narrowness of the apartment not permitting
me conveniently to use a larger size.   I divided
the circumference of this circle into 360 degrees in
the usual manner, and its diameter into thirty
equal parts, which gives about as many minutes as
are equivalent to the Sun's apparent diameter.
Each of these thirty parts was again divided into
four equal portions, making in all one hundred and
twenty ;   and these, if necessary, may be more
minutely subdivided.   The rest I left to ocular
computation, which, in such small sections, is quite
as certain as any mechanical division.   Suppose,

then, each of these thirty parts to be divided into sixty seconds, according to the practice of astronomers. When the time of the observation approached, I retired to my apartment, and, having closed the windows against the light, I directed my telescope—previously adjusted to a focus—through the aperture towards the Sun, and received his rays at right angles upon the paper already mentioned. The Sun's image exactly filled the circle, and I watched carefully and unceasingly for any dark body that might enter upon the disc of light.

'Although the corrected computation of Venus' motions which I had before prepared, and on the accuracy of which I implicitly relied, forbade me to expect anything before three o'clock in the afternoon of the 24th, yet since, according to the calculations of most astronomers, the conjunction should take place sooner—by some even on the 23rd—I was unwilling to depend entirely on my own opinion, which was not sufficiently confirmed, lest by too much self-confidence I might endanger the observation. Anxiously intent, therefore, on the undertaking through the greater part of the 23rd, and on the whole of the 24th, I omitted no available opportunity of observing her ingress. I watched carefully on the 24th from sunrise to nine o'clock, and from a little before ten until noon, and at one in the afternoon, being called away in the intervals by business of the highest importance, which for these ornamental pursuits I could not with propriety neglect.[1] But during all this time I saw nothing in the Sun except a small and common spot, consisting as it were of three points at a distance from the centre towards the left, which I noticed on the preceding and following days. This evidently had nothing to do with Venus. About fifteen minutes past three in the afternoon, when I was again at liberty to continue my labours, the clouds, as if by divine interposition, were entirely dispersed, and I was once more invited to the grateful task of repeating my observations. I then beheld a most agreeable spectacle—the object of my sanguine wishes; a spot of unusual magnitude and of a perfectly circular shape, which had already fully entered upon the Sun's disc on the left, so that the limbs of the Sun and Venus precisely coincided, forming an angle of contact. Not doubting that this was really the shadow of the planet, I immediately applied myself sedulously to observe it.

'In the first place, with respect to the inclination, the line of the diameter of the circle being perpendicular to the horizon, although its plane was somewhat inclined on account of the Sun's altitude, I found that the shadow of Venus at the aforesaid hour—namely, fifteen minutes past three—had entered the Sun's disc about 62° 30', certainly between 60° and 65°, from the top towards the right. This was the appearance in the dark apart-

---

[1] The transit occurred on a Sunday, and the 'business of the highest importance' to which Horrox alludes was his clerical duties.

ment; therefore, out of doors, beneath the open sky, according to the laws of optics, the contrary would be the case, and Venus would be below the centre of the Sun, distant 62° 30' from the lower limbs or the nadir, as the Arabians term it. The inclination remained to all appearances the same until sunset, when the observation was concluded.

'In the second place, the distance between the centres of Venus and the Sun I found by three observations to be as follows :—

The Hour.	Distance of the Centres.
At 3.15 by the clock	14' 24''
„ 3.35 „	13' 30''
„ 3.45 „	13' 0''
„ 3.50 the apparent sun-set.	

The true setting being 3·45, and the apparent about 5 minutes later, the difference being caused by refraction. The clock therefore was sufficiently correct.

'In the third place I found after careful and repeated observation that the diameter of Venus, as her shadow was depicted on the paper, was larger indeed than the thirtieth part of the solar diameter, though not more so than the sixth, or at the utmost the fifth of such a part. Therefore let the diameter of the Sun be to the diameter of Venus as 30' to 1' 12''. Certainly her diameter never equalled 1' 30'', scarcely perhaps 1' 20'', and this was evident as well when the planet was near the Sun's limb as when far distant from it.

'This observation was made in an obscure

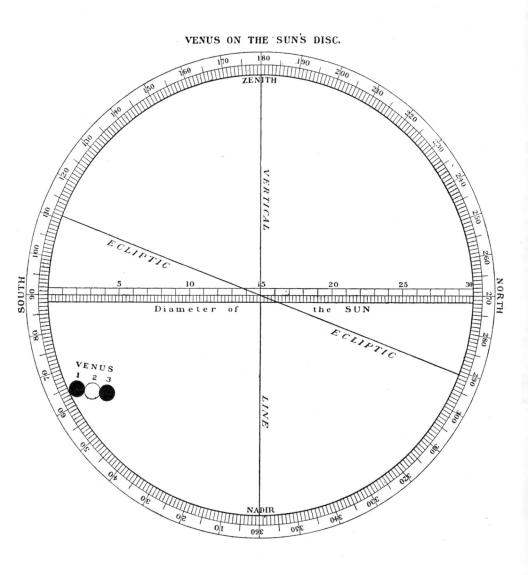

VENUS ON THE SUN'S DISC.

village where I have long been in the habit of
observing, about fifteen miles to the north of Liverpool, the latitude of which I believe to be 53° 20',
although by common maps it is stated at 54° 12',
therefore the latitude of the village will be 53° 35',
and longitude of both 22° 30' from the Fortunate
Islands, now called the Canaries. This is 14° 15'
to the west of Uraniburg in Denmark, the longitude
of which is stated by Brahé, a native of the place,
to be 36° 45' from these islands.

'This is all I could observe respecting this
celebrated conjunction during the short time the
Sun remained in the horizon : for although Venus
continued on his disc for several hours, she was
not visible to me longer than half an hour on
account of his so quickly setting. Nevertheless,
all the observations which could possibly be made
in so short a time I was enabled by Divine Providence to complete so effectually that I could scarcely
have wished for a more extended period. The
inclination was the only point upon which I failed
to attain the utmost precision ; for, owing to the
rapid motion of the Sun it was difficult to observe
with certainty to a single degree, and I frankly
confess that I neither did nor could ascertain it.
But all the rest is sufficiently accurate, and as exact
as I could desire.'

Besides having ascertained that the diameter
of Venus subtends an angle not much greater than
one minute of arc, Horrox reduced the horizontal
solar parallax from fifty-seven seconds as stated by

Kepler to fourteen seconds, a calculation within one and a half second of the value assigned to it by Halley sixty years after. He also reduced the Sun's semi-diameter.

Crabtree, to whom Horrox refers as ‘his most esteemed friend and a person who has few superiors in mathematical learning,’ made preparations to observe the transit similar to those already described. But the day was unfavourable, dark clouds obscured the sky and rendered the Sun invisible. Crabtree was in despair, and relinquished all hope of being able to witness the conjunction. However, just before sunset there was a break in the clouds, and the Sun shone brilliantly for a short interval. Crabtree at once seized his opportunity, and to his intense delight observed the planet fully entered upon the Sun's disc. Instead of proceeding to take observations, he was so overcome with emotion at the sight of the phenomenon, that he continued to gaze upon it with rapt attention, nor did he recover his self-possession until the clouds again hid from his view the setting Sun.[1]

Crabtree's observation of the transit was, however, not a fruitless one. He drew from memory a diagram showing the exact position of Venus on the Sun's disc, which corresponded in every respect with Horrox's observation; he also estimated the diameter of the planet to be $\frac{7}{200}$ that of the Sun, which when calculated gives one minute three

[1] A fresco by the late Mr. Ford Maddox-Brown, depicting Crabtree observing the transit of Venus, adorns the interior of the Manchester Town Hall.

seconds; Horrox having found it to be one minute twelve seconds. This transit of Venus is remarkable as having been the first ever observed of which there is any record, and for this we are indebted to the genius of Horrox, who by a series of calculations, displaying a wonderfully accurate knowledge of mathematics, was enabled to predict the occurrence of the phenomenon on the very day, and almost at the hour it appeared, and of which he and his friend Crabtree were the only observers.

Having thought it desirable to write an account of the transit, Horrox prepared an elegant Latin treatise, entitled 'Venus in Sole Visa '—' Venus seen in the Sun; ' but not knowing what steps to take with regard to its publication, he requested Crabtree to communicate with his bookseller and obtain his advice on the matter.

In the meantime Horrox returned to Toxteth, and arranged to fulfil a long-promised visit to Crabtree, which he looked forward to with much pleasure, as it would afford him an opportunity of discussing with his friend many matters of interest to both. This visit was frustrated in a manner altogether unexpected. For we read that Horrox was seized with a sudden and severe illness, the nature of which is not known, and that his death occurred on the day previous to that of his intended visit to his friend at Broughton. He expired on January 3, 1641, when in the 23rd year of his age.

His death was a great grief to Crabtree, who, in one of his letters, describes it as ' an irreparable loss: ' and it is believed that he only survived him a few years.[1]   Of the papers left by Horrox, only a few have been preserved, and these were discovered in Crabtree's house after his death.   Among them was his treatise on the transit of Venus which, with other papers, was purchased by Dr. Worthington, Fellow of Emmanuel College, Cambridge, a man of learning, who was capable of appreciating their value.   Ultimately, the treatise fell into the possession of Hevelius, a celebrated German astronomer, who published it along with a dissertation of his own, describing a transit of Mercury.

Horrox did not live to see any of his writings published, nor was any monument erected to his memory until nearly two hundred years after his death.   But his name, though long forgotten except by astronomers, is now engraved on marble in Westminster Abbey.   Had his life been spared, it would have been difficult to foretell to what eminence and fame he might have risen, or what further discoveries his genius might have enabled him to make.   Few among English astronomers will hesitate to rank him next with the illustrious Newton, and all will agree with Herschel, who called him ' the pride and the boast of British Astronomy.'

WILLIAM GASCOIGNE was born in 1612, in the parish of Rothwell, in the county of York, and afterwards resided at Middleton, near Leeds.

He was a man of an inventive turn of mind, and possessed good abilities, which he devoted to improving the methods of telescopic observation.

---

[1] William Crabtree died on August 1, 1644, aged 34 years.

At an early age he was occupied in observing celestial objects, making researches in optics, and acquiring a proficient knowledge of astronomy.

Among his acquaintances were Crabtree and Horrox, with whom he carried on a correspondence on matters appertaining to their favourite study.

The measurement of small angles was found at all times to be one of the greatest difficulties which astronomers had to contend with. Tycho Brahé was so misled by his measurements of the apparent diameters of the Sun and Moon, that he concluded a total eclipse of the Sun was impossible.

Gascoigne overcame this difficulty by his invention of the micrometer. This instrument, when applied to a telescope, was found to be of great service in the correct measurement of minute angles and distances, and was the means of greatly advancing the progress of practical astronomy in the seventeenth century. A micrometer consists of a short tube, across the opening of which are stretched two parallel wires ; these being intersected at right angles by a third. The wires are moved to or from each other by delicately constructed screws, to which they are attached. Each revolution, or part of a revolution, of a screw indicates the distance by which the wires are moved.

This apparatus, when placed in the focus of a lens, gives very accurate measurements of the diameters of celestial objects. It was successfully used by Gascoigne in determining the apparent diameters of the Sun, Moon, and several of the planets, and the mutual distances of the stars which form the Pleiades.

Crabtree, after having paid Gascoigne a visit in 1639, describes in a letter to Horrox the impression created on his mind by the micrometer. He writes : ' The first thing Mr. Gascoigne showed me was a large telescope, amplified and adorned with new inventions of his own, whereby he can take the diameters of the Sun or Moon, or any small angle in the heavens or upon the earth, most exactly through the glass to a second.'

The micrometer is now regarded as an indispensable appliance in the observatory ; the use of a spider web reticule instead of wire having improved its efficiency. Gascoigne was one of the earliest astronomers who recognised the value of the Keplerian telescope for observational purposes, and Sherburn affirms that he was the first to construct an instrument of this description having two convex lenses. Whether this be true or not, it is certain that he applied the micrometer to the telescope, and was the first to use telescopic sights, by means of which he was able to fix the optical axis of his telescope, and ascertain by observation the apparent positions of the heavenly bodies.

Crabtree, in a letter to Gascoigne, says : ' Could I purchase it with travel, or procure it with gold, I would not be without a telescope for observing small angles in the heavens ; or want the use of your device of a glass in a cane upon the movable ruler of your sextant, as I remember for helping to the exact point of the Sun's rays.'

It was not known until the beginning of the eighteenth century that Gascoigne had invented and used telescopic sights for the purpose of making accurate astronomical observations. The accidental discovery of some documents which contained a description of his appliances was the means by which this became known.

Townley states that Gascoigne had completed a treatise on optics, which was ready for publication, but that no trace of the manuscript could be discovered after his death. Having embraced the Royalist cause, William Gascoigne joined the forces of Charles I., and fell in the battle of Marston Moor on July 2, 1644.

The early death of this young and remarkably clever man was a severe blow to the science of astronomy in England.

The invention of logarithms, by Baron Napier, of Merchistoun, was found to be of inestimable value to astronomers in facilitating and abbreviating the methods of astronomical calculation.

By the use of logarithms, arithmetical computations which necessitated laborious application for several months could with ease be completed in as many days. It was remarked by Laplace that this invention was the means of doubling the life of an astronomer, besides enabling him to avoid errors and the tediousness associated with long and abstruse calculations.

THOMAS HARRIOT, an eminent mathematician, and an assiduous astronomer, made some valuable observations of the comet of 1607.   He was one of the earliest observers who made use of the telescope, and it was claimed on his behalf that he discovered Jupiter's satellites, and the spots on the Sun, independently of Galileo.   Other astronomers have been desirous of sharing this honour, but it has been conclusively proved that Galileo was the first who made those discoveries.

The investigations of Norwood and Gilbert, the mechanical genius of Hooke, and the patient researches of Flamsteed—the first Astronomer Royal —were of much value in perfecting many details associated with the study of astronomy.

The Royal Observatory at Greenwich was founded in 1675.   The building was erected under a warrant from Charles II.   It announces the desire of the Sovereign to build a small observatory in the park at Greenwich, ' in order to the finding out of the longitude for perfecting the art of navigation and astronomy.'   This action on the part of the King may be regarded as the first public acknowledgment of the usefulness of astronomy for national purposes.

Since its erection, the observatory has been presided over by a succession of talented men, who have raised it to a position of eminence and usefulness unsurpassed by any similar institution in this or any other country.   The well-known names of Flamsteed, Halley, Bradley, and Airy, testify to

the valuable services rendered by those past direc-
tors of the Greenwich Observatory in the cause of
astronomical science.

If we take a general survey of the science of
astronomy as it existed from 1608 to 1674—a period
that embraced the time in which Milton lived—we
shall find that it was still compassed by ignorance,
superstition, and mystery. Astrology was zealously
cultivated ; most persons of rank and position had
their nativity or horoscope cast, and the belief in
the ruling of the planets, and their influence on
human and terrestrial affairs, was through long
usage firmly establish in the public mind. Indeed,
at this time, astronomy was regarded as a hand-
maid to astrology ; for, with the aid of astronomical
calculation, the professors of this occult science were
enabled to predict the positions of the planets, and
by this means practised their art with an apparent
degree of truthfulness.

Although over one hundred years had elapsed
since the death of Copernicus, his theory of the
solar system did not find many supporters, and the
old forms of astronomical belief still retained their
hold on the minds of the majority of philosophic
thinkers. This can be partly accounted for, as many
of the Ptolemaic doctrines were at first associated
with the Copernican theory, nor was it until a later
period that they were eliminated from the system.

Though Copernicus deserved the credit of having
transferred the centre of our system from the Earth
to the Sun, yet his theory was imperfect in its

details, and contained many inaccuracies. He
believed that the planets could only move round
the Sun in circular paths, nor was he capable of
conceiving of any other form of orbit in which
they could perform their revolutions. He was
therefore compelled to retain the use of cycles
and epicycles, in order to account for irregularities in the uniformly circular motions of those
bodies.

We are indebted to the genius of Kepler for
having placed the Copernican system upon a sure
and irremovable basis, and for having raised astronomy to the position of a true physical science. By
his discovery that the planets travel round the Sun
in elliptical orbits, he was enabled to abolish cycles
and epicycles, which created such confusion and
entanglement in the system, and to explain many
apparent irregularities of motion by ascribing to the
Sun his true position with regard to the motions of
the planets.

After the death of Kepler, which occurred in
1630, the most eminent supporter of the Copernican
theory was the illustrious Galileo, whose belief in
its accuracy and truthfulness was confirmed by his
own discoveries.

Five of the planets were known at this time—viz.
Mercury, Venus, Mars, Jupiter, and Saturn; the
latter, which revolves in its orbit at a profound
distance from the Sun, formed what at that time
was believed to be the boundary of the planetary
system. The distance of the Earth from the Sun

was approximately known, and the orb was observed to rotate on his axis.

It was also ascertained that the Moon shone by reflected light, and that her surface was varied by inequalities resembling those of our Earth. The elliptical form of her orbit had been discovered by Horrox, and her elements were computed with a certain degree of accuracy.

The cloudy luminosity of the Milky Way had been resolved into a multitude of separate stars, disclosing the immensity of the stellar universe.

The crescent form of the planet Venus, the satellites of Jupiter and of Saturn, and the progressive motion and measurement of light, had also been discovered. Observations were made of transits of Mercury and Venus, and refracting and reflecting telescopes were invented.

The law of universal gravitation, a power which retains the Earth and planets in their orbits, causing them year after year to describe with unerring regularity their oval paths round the Sun, was not known at this time. Though Newton was born in 1642, he did not disclose the results of his philosophic investigations until 1687—thirteen years after the death of Milton—when, in the ' Principia,' he announced his discovery of the great law of universal gravitation.

Kepler, though he discovered the laws of planetary motion, was unable to determine the motive force which guided and retained those bodies in their orbits. It was reserved for the genius of Newton to solve this wonderful problem.   This great philosopher was able to prove 'that every particle of matter in the universe attracts every other particle with a force proportioned to the mass of the attracting body, and inversely as the square of the distance between them.'   Newton was capable of demonstrating that the force which guides and retains the Earth and planets in their orbits resides in the Sun, and by the application of this law of gravitation he was able to explain the motions of all celestial bodies entering into the structure of the solar system.

This discovery may be regarded as the crowning point of the science of astronomy, for, upon the unfailing energy of this mysterious power depend the order and stability of the universe, extending as it does to all material bodies existing in space, guiding, controlling, and retaining them in their several paths and orbits, whether it be a tiny meteor, a circling planet, or a mighty sun.

The nature of cometary bodies and the laws which govern their motions were at this time still enshrouded in mystery, and when one of those erratic wanderers made its appearance in the sky it was beheld by the majority of mankind with feelings of awe and superstitious dread, and regarded as a harbinger of evil and disaster, the precursor of war, of famine, or the overthrow of an empire.

Newton, however, was able to divest those bodies of the mystery with which they were surrounded by proving that any conic section may be described about the Sun, consistent with the law of gravitation, and that comets, notwithstanding the eccentricity of their orbits, obey the laws of planetary motion.

Beyond the confines of our solar system, little was known of the magnitude and extent of the sidereal universe which occupies the infinitude of space by which we are surrounded. The stars were recognised as self-luminous bodies, inconceivably remote, and although they excited the curiosity of observers, and conjectures were made as to their origin, yet no conclusive opinions were arrived at with regard to their nature and constitution, and except that they were regarded as glittering points of light which illumine the firmament, all else appertaining to them remained an unravelled mystery. Even Copernicus had no notion of a universe of stars.

Galileo, by his discovery that the galaxy consists of a multitude of separate stars too remote to be defined by ordinary vision, demonstrated how vast are the dimensions of the starry heavens, and on what a stupendous scale the universe is constructed. But at this time it had not occurred to astronomers, nor was it known until many years after, that the stars are suns which shine with a splendour resembling that of our Sun, and in many instances surpassing it. It was not until this truth became known that the glories of the sidereal heavens were fully comprehended, and their magnificence revealed.

It was then ascertained that the minute points of
light which crowd the fields of our largest telescopes,
in their aggregations forming systems, clusters,
galaxies, and universes of stars, are shining orbs of
light, among the countless multitudes of which our
Sun may be numbered as one.

CHAPTER III

MILTON'S ASTRONOMICAL KNOWLEDGE

IT would be reasonable to imagine that Milton's knowledge of astronomy was comprehensive and accurate, and superior to that possessed by most scientific men of his age. His scholarly attainments, his familiarity with ancient history and philosophy, his profound learning, and the universality of his general knowledge, would lead one to conclude that the science which treats of the mechanism of the heavens, and especially the observational part of it—which at all times has been a source of inspiration to poets of every degree of excellence—was to him a study of absorbing interest, and one calculated to make a deep impression upon his devoutly poetical mind. The serious character of Milton's verse, and the reverent manner in which celestial incidents and objects are described in it, impress one with the belief that his contemplation of the heavens, and of the orbs that roll and shine in the firmament overhead, afforded him much enjoyment and meditative delight. For no poet, in ancient or in modern times, has introduced into his writings with such frequency, or with such pleasing effect, so many passages descriptive

G

of the beauty and grandeur of the heavens. No other poet, by the creative effort of his imagination, has soared to such a height; nor has he ever been excelled in his descriptions of the celestial orbs, and of the beautiful phenomena associated with their different motions.

In his minor poems, which were composed during his residence at Horton, a charming rural retreat in Buckinghamshire, where the freshness and varied beauty of the landscape and the attractive aspects of the midnight sky were ever before him, we find enchanting descriptions of celestial objects, and especially of those orbs which, by their brilliancy and lustre, have always commanded the admiration of mankind.

For example, in 'L'Allegro' there are the following lines :—

> Right against the eastern gate
> Where the great Sun begins his state,
> Robed in flames and amber light,
> The clouds in thousand liveries dight;

and in 'Il Penseroso '—

> To behold the wandering Moon,
> Riding near her highest noon,
> Like one that had been led astray
> Through the heaven's wide pathless way,
> And oft as if her head she bowed,
> Stooping through a fleecy cloud.

In the happy choice of his theme, and by the comprehensive manner in which he has treated it, Milton has been enabled by his poetic genius to give to the world in his 'Paradise Lost' a poem which, for sublimity of thought, loftiness of imagination, and beauty of expression in metrical verse, is unsurpassed in any language.

It is, however, our intention to deal only with those passages in the poem in which allusion is made to the heavenly bodies, and to incidents and occurrences associated with astronomical phenomena.   In the exposition and illustration of these it has been considered desirable to adopt the following general classification :—

1. To ascertain the extent of Milton's astronomical knowledge.

2. To describe the starry heavens and the celestial objects mentioned in ' Paradise Lost.'

3. To exemplify the use which Milton has made of astronomy in the exercise of his imaginative and descriptive powers.

In the earlier half of the seventeenth century the Ptolemaic theory—by which it was believed that the Earth was the immovable centre of the universe, and that round it all the heavenly bodies completed a diurnal revolution—still retained its ascendency over the minds of men of learning and science, and all the doctrines associated with this ancient astronomical creed were still religiously upheld by the educated classes among the peoples inhabiting the different civilised regions of the globe. The Copernican theory—by which the Sun is assigned the central position in our system, with the Earth and planets revolving in orbits round him—obtained the support of a few persons of advanced views and high scientific attainments, but its doctrines had not yet seriously threatened the supremacy of the older system. Though upwards of one hundred years had elapsed since the death of Copernicus, yet the doctrines associated with the system of which he was the founder were but very tardily adopted up to this time. There were several reasons which accounted for this. The Copernican system was at first imperfect in its details, and included several of the Ptolemaic, doctrines which rendered it less intelligible, and retarded its acceptance by persons who would otherwise have been inclined to adopt it. Copernicus believed that the planets travelled round the Sun in circular paths. This necessitated the retention of cycles and epicycles, which gave rise to much confusion; nor was it until Kepler made his great discovery of the ellipticity of the planetary orbits that they were eliminated from the system.

As the Ptolemaic system of the universe held complete sway over the minds of men for upwards of twenty centuries, it was difficult to persuade many persons to renounce the astronomical beliefs to which they were so firmly attached, in favour of those of any other system; so that the overthrow of this venerable theory required a lengthened period of time for its accomplishment.

It was thus in his earlier years, when Milton devoted his time to the study of literature and philosophy, which he read extensively when pursuing his academic career at Christ's College, Cam-

bridge, and afterwards at Horton, where he spent
several years in acquiring a more proficient knowledge of the literary, scientific, and philosophical
writings of the age, that he found the beliefs associated with the Ptolemaic theory adopted without
doubt or hesitation by the numerous authors whose
works he perused.   His knowledge of Italian
enabled him to become familiar with Dante—one
of his favourite authors, whose poetical writings
were deeply read by him, and who, in the elaboration of his poem, the ' Divina Commedia,' included
the entire Ptolemaic cosmology.

In England the Copernican theory had few
supporters, and the majority of those who represented the intellect and learning of the country
still retained their adherence to the old form of
astronomical belief.   We therefore find that Milton
followed the traditional way of thinking by adopting
the views associated with the Ptolemaic theory.

According to the Ptolemaic system, the Earth
was regarded as the immovable centre of the universe, and surrounding it were ten crystalline
spheres, or heavens, arranged in concentric circles,
the larger spheres enclosing the smaller ones ; and
within those was situated the cosmos, or mundane
universe, usually described as ' the Heavens and the
Earth.'   To each of the first seven spheres there
was attached a heavenly body, which was carried
round the Earth by the revolution of the crystalline.

1st sphere : that of the Moon.

2nd sphere : that of the planet Mercury.

3rd sphere : that of the planet Venus.

4th sphere : that of the Sun ; regarded as a planet.

5th sphere : that of the planet Mars.

6th sphere : that of the planet Jupiter.

7th sphere : that of the planet Saturn.

8th sphere : that of the fixed stars.

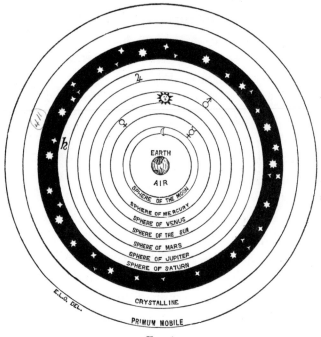

Fig. 1

The eighth sphere included all the fixed stars, and was called the firmament, because it was believed to impart steadiness to the inner spheres, and, by its diurnal revolution, to carry them round the Earth, causing the change of day and night.

The separate motions of the spheres, revolving with different velocities, and at different angles to each other, accounted for the astronomical phenomena associated with the orbs attached to each. According to Ptolemy's scheme, the eighth sphere formed the outermost boundary of the universe; but later astronomers added to this system two other spheres—a *ninth*, called the *Crystalline*, which caused Precession of the Equinoxes; and a *tenth*, called the *Primum Mobile*, or First Moved, which brought about the alternation of day and night, by carrying all the other spheres round the Earth once in every twenty-four hours. The Primum Mobile enclosed, as if in a shell, all the other spheres, in which was included the created universe, and, although of vast dimensions, its conception did not overwhelm the mind in the same manner that the effort to comprehend infinitude does.

Beyond this last sphere there was believed to exist a boundless, uncircumscribed region, of immeasurable extent, called the Empyrean, or Heaven of Heavens, the incorruptible abode of the Deity, the place of eternal mysteries, which the comprehension of man was unable to fathom, and of which it was impossible for his mind to form any conception. Such were the imaginative beliefs upon which this ancient astronomical theory was founded, that for a period of upwards of two thousand years held undisputed sway over the minds of men, and exercised during that time a predominating influence upon the imagination, thoughts, and concep-

tions of all those who devoted themselves to literature,
science, and art.   Of the truthfulness of this asser-
tion there is ample evidence in the poetical, philosophical, and historical writings of ancient authors,
whose ideas and conceptions regarding the created
universe were limited and circumscribed by this
form of astronomical belief.   In the works of more
recent writers we find that it continued to assert
its influence; and among our English poets, from
Chaucer down to Shakespeare, there are numerous
references to the natural phenomena associated
with this system, and most frequently expressed by
poetical allusions to ' the music of the spheres.'

The ideas associated with the Ptolemaic theory
were gratifying to the pride and vanity of man, who
could regard with complacency the paramount
importance of the globe which he inhabited, and
of which he was the absolute ruler, fixed in the
centre of the universe, and surrounded by ten
revolving spheres, that carried along with them in
their circuit all other celestial bodies—Sun, Moon,
and stars, which would appear to have been created
for his delectation, and for the purpose of minister-
ing to his requirements.   But when the Copernican
theory became better understood, and especially
after the discovery of the law of universal gravitation, this venerable system of the universe, based
upon a pile of unreasonable and false hypotheses,
after an existence of over twenty centuries, sank
into oblivion, and was no more heard of.

Milton's Ptolemaism is apparent in some of his

shorter pieces, and also in his minor poems, 'Arcades' and 'Comus.' His 'Ode on the Nativity' is written in conformity with this belief, and the expression,

Ring out ye crystal spheres,

indicates a poetical allusion to this theory. But as Milton grew older his Ptolemaism became greatly modified, and there are good reasons for believing that in his latter years he renounced it entirely in favour of Copernicanism. When on his continental tour in 1638, he made the acquaintance of eminent men who held views different from those with which he was familiar ; and in his interview with Galileo at Arcetri, the aged astronomer may have impressed upon his mind the superiority of the Copernican theory, in accounting for the occurrence of celestial phenomena, as compared with the Ptolemaic.

On his return to England from the Continent, Milton took up his residence in London, and lived in apartments in a house in St. Bride's Churchyard. Having no regular vocation, and not wishing to be dependent upon his father, he undertook the education of his two nephews, John and Edward Phillips, aged nine and ten years respectively. From St. Bride's Churchyard he removed to a larger house in Aldersgate, where he received as pupils the sons of some of his most intimate acquaintances. In the list of subjects which Milton selected for the purpose of imparting instruction to those youths he included astronomy and mathematics, which formed part of the curriculum of this educational establishment. The text-book from which he taught his nephews and other pupils astronomy was called ' De Sphæra Mundi,' a work written by Joannes Sacrobasco (John Holywood) in the thirteenth century. This book was an epitome of Ptolemy's 'Almagest,' and therefore entirely Ptolemaic in its teaching. It enjoyed great popularity during the Middle Ages, and is reported to have gone through as many as forty editions.

The selection of astronomy as one of the subjects in which Milton instructed his pupils affords us evidence that he must have devoted considerable time and attention to acquiring a knowledge of the facts and details associated with the study of the science. In the attainment of this he had to depend upon his own exertions and the assistance derived from astronomical books; for at this time astronomy received no recognition as a branch of study at any of the universities ; and in Britain the science attracted less attention than on the Continent, where the genius of Kepler and Galileo elevated it to a position of national importance.

We shall find as we proceed that Milton's knowledge of astronomy was comprehensive and accurate; that he was familiar with the astronomical reasons by which many natural phenomena which occur around us can be explained ; and that he understood many of the details of the science which are unknown to ordinary observers of the heavens.

It is remarkable how largely astronomy enters into the composition of 'Paradise Lost,' and we doubt if any author could have written such a poem without possessing a knowledge of the heavens and of the celestial orbs such as can only be attained by a proficient and intimate acquaintance with this science.

The arguments in favour of or against the Ptolemaic and Copernican theories were well known to Milton, even as regards their minute details; and in Book viii. he introduces a scientific discussion based upon the respective merits of those theories. The configuration of the celestial and terrestrial spheres, and the great circles by which they are circumscribed, he also knew. The causes which bring about the change of the seasons; the obliquity of the ecliptic; the zodiacal constellations through which the Sun travels, and the periods of the year in which he occupies them, are embraced in Milton's knowledge of the science of astronomy. The motions of the Earth, including the Precession of the Equinoxes; the number and distinctive appearances of the planets, their direct and retrograde courses, and their satellites, are also described by him. The constellations, and their relative positions on the celestial sphere; the principal stars, star-groups, and clusters, and the Galaxy, testify to Milton's knowledge of astronomy, and to the use which he has made of the science in the elaboration of his poem.

The names of fourteen of the constellations are

mentioned in 'Paradise Lost.' These, when arranged alphabetically, read as follows :—

Andromeda, Aries, Astrea, Centaurus, Cancer,
Capricornus, Gemini, Leo, Libra, Ophiuchus, Orion,
Scorpio, Taurus, and Virgo. Milton's allusions to
the zodiacal constellations are chiefly associated
with his description of the Sun's path in the heavens;
but with the celestial sign Libra (the *Scales*) he
has introduced a lofty and poetical conception of
the means by which the Creator made known
His will when there arose a contention between
Gabriel and Satan on his discovery in Paradise.

> The Eternal, to prevent such horrid fray,
> Hung forth in Heaven his golden scales, yet seen
> Betwixt Astrea [1] and the Scorpion sign,
> Wherein all things created first he weighed,
> The pendulous round Earth with balanced air
> In counterpoise, now ponders all events,
> Battles and realms. In these he put two weights,
> The sequel each of parting and of fight:
> The latter quick up flew, and kicked the beam.
>
> —iv. 996–1004.

Orion, the finest constellation in the heavens,
did not escape Milton's observation, and there is
one allusion to it in his poem. It arrives on the
meridian in winter, where it is conspicuous as a
brilliant assemblage of stars, and represents an
armed giant, or hunter, holding a massive club in
his right hand, and having a shield of lion's hide
on his left arm. A triple-gemmed belt encircles
his waist, from which is suspended a glittering

---

[1] The constellation Virgo.

sword, tipped with a bright star.   The two brilliants
Betelgeux and Bellatrix form the giant's shoulders,
and the bright star Rigel marks the position of his
advanced foot.   The rising of Orion was believed
to be accompanied by stormy and tempestuous
weather.   Milton alludes to this in the following
lines :—

> When with fierce winds Orion armed
> Hath vexed the Red Sea coast, whose waves o'erthrew
> Busiris and his Memphian chivalry.—i. 305-7.

Andromeda is described as being borne by Aries,
and in 'Ophiuchus huge' Milton locates a comet
which extends the whole length of the constellation.
It is evident that Milton possessed a precise knowledge of the configuration and size of the constellations, and of the positions which they occupy
relatively to each other on the celestial sphere.

Though Milton was conversant with the Copernican theory, and entertained a conviction of its
accuracy and truthfulness, and doubtless recognised
the superiority of this system, which, besides con-
veying to the mind a nobler conception of the universe and of the solar system—though it diminished
the importance of the Earth as a member of it—
was capable of explaining the occurrence of celestial
phenomena in a manner more satisfactory than
could be arrived at by the Ptolemaic theory.   Notwithstanding this, he selected the Ptolemaic cosmology as the scientific basis upon which he constructed
his 'Paradise Lost,' and in its elaboration adhered
with marked fidelity to this system.   There were

many reasons why Milton, in the composition of an imaginative poem, should have chosen the Ptolemaic system of the universe rather than the Copernican. This form of astronomical belief was adopted by all the authors whose works he perused and studied in his younger days, including his favourite poet, Dante; and his own poetic imaginings, as indicated by his early poems, were in harmony with the doctrines of this astronomical creed, a long acquaintance with which had, without doubt, influenced his mind in its favour. This system of revolving spheres, with the steadfast Earth at its centre, and the whole enclosed by the Primum Mobile, constituted a more attractive and picturesque object for poetic description than the simple and uncircumscribed arrangement of the universe expressed by the Copernican theory. It also afforded him an opportunity of localising those regions of space in which the chief incidents in his poem are described—viz. HEAVEN, or THE EMPYREAN, CHAOS, HELL, and the MUNDANE UNIVERSE. Milton's Ptolemaism, with its adjuncts, may be understood by the following :

All that portion of space above the newly created universe, and beyond the Primum Mobile, was known as HEAVEN, or THE EMPYREAN—a region of light, of glory, and of happiness ; the dwelling-place of the Deity, Who, though omnipresent, here visibly revealed Himself to all the multitude of angels whom He created, and who surrounded his throne in adoration and worship.

Underneath the universe there existed a vast region of similar dimensions to the Empyrean, called CHAOS, which was occupied by the embryo elements of matter, that with incessant turmoil and confusion warred with each other for supremacy—a wild abyss—

The womb of Nature and perhaps her grave.—ii. 911.

The lower portion of this region was divided off from the remainder, and embraced the locality known as HELL—the place of torment, where the rebellious angels were driven and shut in after their expulsion from Heaven.

As far removed from God and light of Heaven
As from the centre thrice to the utmost pole.—i. 73-74.

The NEW UNIVERSE, which included the Earth and all the orbs of the firmament known as the Starry Heavens, was created out of Chaos, and hung, as if suspended by a golden chain, from the Empyrean above; and although its magnitude and dimensions were inconceivable, yet, according to the Ptolemaic theory, it was enclosed by the tenth sphere or Primum Mobile.

By this partitioning of space Milton was able to contrive a system which fulfilled the requirements of his great poem.

The annexed diagram explains the relative positions of the different regions into which space was divided.

Though there are traces of Copernicanism found in 'Paradise Lost,' yet Milton has very faithfully adhered to the Ptolemaic mechanism and nomenclature throughout his poem.

In his description of the Creation, the Earth is formed first, then the Sun, followed by the Moon, and afterwards the stars, all of which are described as being in motion round the Earth. Allusion is also made to this ancient system in several pro-

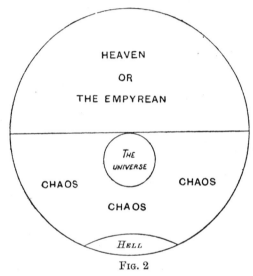

Fig. 2

minent passages, and in the following lines there is a distinct reference to the various revolving spheres.

> They pass the planets seven, and pass the fixed,
> And that crystalline sphere whose balance weighs
> The trepidation talked, and that first moved.
> —iii. 481–83.

The seven planetary spheres are first mentioned; then the eighth sphere, or that of the fixed stars; then the ninth, or crystalline, which was believed to cause a shaking, or trepidation, to account for certain irregularities in the motions of the stars ; and, lastly, the tenth sphere, or Primum Mobile, called the ' first moved' because it set the other spheres in motion.

To an uninstructed observer, the apparent motion of the heavenly bodies round the Earth would naturally lead him to conclude that, of the two theories, the Ptolemaic was the correct one. We therefore find that Milton adopted the system most in accord with the knowledge and intelligence possessed by the persons portrayed by him in his poem ; and in describing the natural phenomena witnessed in the heavens by our first parents, he adheres to the doctrines of the Ptolemaic system, as being most in harmony with the simple and primitive conceptions of those created beings.

To their upward gaze, the orbs of heaven appeared to be in ceaseless motion ; the solid Earth, upon which they stood, was alone immovable and at rest. Day after day they observed the Sun pursue his steadfast course with unerring regularity : his rising in the east, accompanied by the rosy hues of morn ; his meridian splendour, and his sinking in the west, tinting in colours of purple and gold inimitable the fleecy clouds floating in the azure sky, as he bids farewell for a time to scenes of life and happiness, rejoicing in the light and warmth of his all-cheering beams. With the advent of night they beheld the Moon, now increasing, now waning, pursue her irregular path, also to

H

disappear in the west; whilst, like the bands of an army marshalled in loose array, the constellations of glittering stars, with stately motion, traversed their nocturnal arcs, circling the pole of the heavens.

By referring to Book viii., 15–175, we find an account of an interesting scientific discussion, or conversation, between Adam and Raphael regarding the merits of the Ptolemaic and Copernican systems, and of the relative importance and size of the heavenly bodies. By it we are afforded an opportunity of learning how accurate and precise a knowledge Milton possessed of both theories, and in what clear and perspicuous language he expresses his arguments in favour of or against the doctrines associated with each.

We may, with good reason, regard the views expressed by Adam as representing Milton's own opinions, which were in conformity with the Copernican theory; and in the Angel's reply, though of an undecided character, we are able to perceive how aptly Milton describes the erroneous conclusions upon which the Ptolemaic theory was based.

In this scientific discussion, it would seem rather strange that Adam, the first of men, should have been capable of such philosophic reasoning, propounding, as if by intuition, a theory upon which was founded a system that had not been discovered until many centuries after the time that astronomy became a science. By attributing to Adam such a degree of intelligence and wisdom, the poet has taken a liberty which enabled him to carry on this discussion in a manner befitting the importance of the subject.

In the following lines Adam expresses to his Angel-guest, in forcible and convincing language, his reasons in support of the Copernican theory :—

> When I behold this goodly frame, this World,
> Of Heaven and Earth consisting, and compute
> Their magnitudes—this Earth, a spot, a grain,
> An atom, with the Firmanent compared
> And all her numbered stars, that seem to roll
> Spaces incomprehensible (for such
> Their distance argues, and their swift return
> Diurnal) merely to officiate light
> Round this opacous Earth, this punctual spot,
> One day and night, in all her vast survey
> Useless besides—reasoning, I oft admire,
> How Nature, wise and frugal could commit
> Such disproportions, with superfluous hand
> So many nobler bodies to create,
> Greater so manifold, to this one use,
> For aught appears, and on their Orbs impose
> Such restless revolution day by day
> Repeated, while the sedentary Earth,
> That better might with far less compass move,
> Served by more noble than herself, attains
> Her end without least motion, and receives,
> As tribute, such a sumless journey brought
> Of incorporeal speed, her warmth and light ;
> Speed, to describe whose swiftness number fails.
>
> viii. 15-38.

We are enabled to perceive that Milton had formed a correct conception of the magnitude and proportions of the universe, and also of the relative size and importance of the Earth, which he describes as ' a spot, a grain, an atom,' when compared

with the surrounding heavens. He expresses his
surprise that all the stars of the firmament, whose
distances are so remote, and whose dimensions so
greatly exceed those of this globe, should in their
diurnal revolution have 'such a sumless journey of
incorporeal speed imposed upon them' merely to
officiate light to the Earth, 'this punctual spot;'
and reasoning, wonders how Nature, wise and
frugal in her ways, should commit such dispropor-
tions, by adopting means so great to accomplish a
result so small, when motion imparted to the
sedentary Earth would with greater ease produce
the same effect.

The inconceivable velocity with which it would
be necessary for those orbs to travel in order to
accomplish a daily revolution round the Earth
might be described as almost spiritual, and beyond
the power of calculation by numbers.

The Angel, after listening to Adam's argument,
expresses approval of his desire to obtain knowledge, but answers him dubiously, and at the same
time criticises in a severe and adverse manner the
Ptolemaic theory.

> To ask or search I blame thee not; for Heaven
> Is as the Book of God before thee set,
> Wherein to read his wondrous works, and learn
> His seasons, hours, or days, or months, or years.
> This to attain, whether Heaven move or Earth,
> Imports not, if thou reckon right; the rest
> From Man or Angel the Great Architect
> Did wisely to conceal, and not divulge
> His secrets, to be scanned by them who ought
> Rather admire.   Or, if they list to try

Conjecture, he his fabric of the Heavens
Hath left to their disputes, perhaps to move
His laughter at their quaint opinions wide
Hereafter, when they come to model Heaven,
And calculate the stars ; how they will wield
The mighty frame ; how build, unbuild, contrive
To save appearances ; how gird the Sphere
With Centric and Eccentric scribbled o'er
Cycle and Epicycle, Orb in Orb.—viii. 66–84.

When, with the advancement of science, astronomical observations were made with greater accuracy, it was discovered that uniformity of motion was
not always maintained by those bodies which were
believed to move in circles round the Earth. It
was observed that the Sun, when on one side of his
orbit, had an accelerated motion, as compared with
the speed at which he travelled when on the other
side. The planets, also, appeared to move with
irregularity : sometimes a planet was observed to
advance, then become stationary, and afterwards
affect a retrograde movement. Those inequalities
of motion could not be explained by means of the
revolution of crystalline spheres alone, but were
accounted for by imagining the existence of a small
circle, or epicycle, whose centre corresponded with
a fixed point in the larger circle, or eccentric, as it
was called. This small circle revolved on its axis
when carried round with the larger one, and round
it the planet also revolved, which when situated in
its outer portion would have a forward, and when
in its inner portion a retrograde, motion.

The theory of eccentrics and epicycles was suffi-

cient for a time to account for the inequalities of
motion already described, and by this means the
Ptolemaic system was enabled to retain its ascendency for a longer period than it otherwise would
have done.   But more recent discoveries brought
to light discrepancies and difficulties which were
explained away by adding epicycle to epicycle.   This
created a most complicated entanglement, and
hastened the downfall of a system which, after an
existence of many centuries, sank into oblivion, and
is now remembered as a belief of bygone ages.

The devices which the upholders of this system
were compelled to adopt, in order ' to save appearances,' with ' centric and eccentric,' cycle and epicycle, ' orb in orb,' are in this manner appropriately
described by Milton, as indicating the confusion
arising from a theory based upon false hypotheses.
Continuing his reply, the Angel says :—

> Already by thy reasoning this I guess,
> Who art to lead thy offspring, and supposest
> That bodies bright and greater should not serve
> The less not bright, nor Heaven such journies run,
> Earth sitting still, when she alone receives
> The benefit.   Consider, first, that great
> Or bright infers not excellence.   The Earth,
> Though, in comparison of Heaven, so small,
> Nor glistering, may of solid good contain
> More plenty than the Sun that barren shines,
> Whose virtue on itself works no effect,
> But in the fruitful Earth ; there first received,
> His beams, inactive else, their vigour find,
> Yet not to Earth are those bright luminaries
> Officious, but to thee, Earth's habitant.
> And, for the Heaven's wide circuit, let it speak

The Maker's high magnificence, who built
So spacious, and his line stretched out so far,
That Man may know he dwells not in his own—
An edifice too large for him to fill,
Lodged in a small partition ; and the rest
Ordained for uses to his Lord best known,
The swiftness of those Circles attribute,
Though numberless, to his Omnipotence,
That to corporeal substances could add
Speed almost spiritual. Me thou think'st not slow,
Who since the morning-hour set out from Heaven
Where God resides, and ere midday arrived
In Eden—distance inexpressible
By numbers that have name. But this I urge,
Admitting motion in the Heavens, to show
Invalid that which thee to doubt it moved ;
Not that I so affirm, though so it seem
To thee who hast thy dwelling here on Earth.
God, to remove his ways from human sense,
Placed Heaven from Earth so far, that earthly sight,
If it presume, might err in things too high,
And no advantage gain.—viii. 85–122.

Notwithstanding the Angel's severe criticism of
the Ptolemaic system, he does not unreservedly
support the conclusions arrived at by Adam, but
endeavours to show that his reasoning may not be
altogether correct. He questions the validity of his
argument that bodies of greater size and brightness should not serve the smaller, though not bright,
and that heaven should move, while the Earth
remained at rest. He argues that great or bright
infers not excellence, and that the Earth, though
small, may contain more virtue than the Sun, that
' barren shines,' whose beams create no beneficial
effect, except when directed on the fruitful Earth.

He reminds Adam that those bright luminaries minister not to the Earth, but to himself, 'Earth's habitant,' and directs his attention to the magnificence and extent of the surrounding universe, of which he occupies but a small portion. The diurnal swiftness of the orbs that move round the Earth he attributes to God's omnipotence, that to material bodies 'could add speed almost spiritual.'

The Angel, after alluding to his rapid flight through space, suggests that God placed heaven so far from Earth that man might not presume to inquire into things which it would be of no advantage for him to know. He then suddenly changes to the Copernican system, which he lucidly describes in the following lines :—

> What if the Sun
> Be centre to the World, and other stars
> By his attractive virtue and their own
> Incited, dance about him various rounds ?
> Their wandering course, now high, now low, then hid,
> Progressive, retrograde, or standing still,
> In six thou seest ; and what if, seventh to these
> The planet Earth, so steadfast though she seem,
> Insensibly three different motions move ?
> Which else to several spheres thou must ascribe,
> Moved contrary with thwart obliquities,
> Or save the Sun his labour, and that swift
> Nocturnal and diurnal rhomb supposed
> Invisible else above all stars, the wheel
> Of day and night ; which needs not thy belief,
> If Earth, industrious of herself, fetch day
> Travelling east, and with her part averse
> From the Sun's beam meet night, her other part
> Still luminous by his ray. What if that light,
> Sent from her through the wide transpicuous air,

To the terrestrial Moon be as a star,
Enlightening her by day, as she by night
This Earth—reciprocal, if land be there,
Fields and inhabitants ?   Her spots thou seest
As clouds, and clouds may rain, and rain produce
Fruits in her softened soil, for some to eat
Allotted there ; and other Suns, perhaps,
With their attendant Moons, thou wilt descry,
Communicating male and female light—
Which two great sexes animate the World,
Stored in each orb perhaps with some that live.
For such vast room in Nature unpossessed
By living soul, desert and desolate,
Only to shine, yet scarce to contribute
Each orb a glimpse of light, conveyed so far
Down to this habitable, which returns
Light back to them, is obvious to dispute.—viii. 122–58.

The Copernican theory, which is less complicated
and more easily understood than the Ptolemaic,
is described by Milton with accuracy and methodical
skill.

The Sun having been assigned that central position in the system which his magnitude and importance claim as his due, the planets circling in orbits
around him have their motions described in a manner indicative of the precise knowledge which Milton
acquired of this theory.   At this time the law of
gravitation was unknown, and, although the ellipticity of the orbits of the planets had been discovered
by Kepler, the nature of the motive force which
guided and retained them in their paths still remained a mystery.   It was believed that the planets
were whirled round the Sun, as if by the action of
magnetic fibres ; a mutual attractive influence

having been supposed to exist between them and the orb, similar to that of the opposite poles of magnets.

Milton alludes to this theory in the following lines :—

> They, as they move
> Their starry dance in numbers that compute
> Days, months, and years, towards his all-cheering lamp
> Turn swift their various motions, or are turned
> By his magnetic beam.—iii. 579–83.

An important advance upon this theory was made by Horrox, who, in his study of celestial dynamics, attributed the curvilineal motion of the planets to the influence of two forces, one projective, the other attractive. He illustrated this by observing the path described by a stone when thrown obliquely into the air. He perceived that its motion was governed by the impulse imparted to it by the hand, and also by the attractive force of the Earth. Under these two influences, the stone describes a graceful curve, and in its descent falls at the same angle at which it rose. Hence arises the general law : 'When two spheres are mutually attracted, and if not prevented by foreign influences, their straight paths are deflected into curves concave to each other, and corresponding with one of the sections of a cone, according to the velocity of the revolving body. If the velocity with which the revolving body is impelled be equal to what it would acquire by falling through half the radius of a circle described from the centre of deflection, its orbit will be circular ; but if it be less than that quantity, its path becomes elliptical.'

Newton afterwards embraced this law in his great principle of gravitation, and demonstrated that the force which guides and retains the Earth and planets in their orbits resides in the Sun. By the orb's attractive influence a planet, after having received its first impulse, is deflected from its original straight path, and bent towards that luminary, and by the combined action of the projective and attractive forces is made to describe an orbit which, if elliptical, has one of its foci occupied by the Sun. So evenly balanced are those two forces, that one is unable to gain any permanent ascendency over the other, and consequently the planet traverses its orbit with unerring regularity, and, if undisturbed by external influences, will continue in its path for all time.

Milton describes the position of the planets in the sky as—

> Now high, now low, then hid;

and their motions—

> Progressive, retrograde, or standing still.

It is evident that Milton was familiar with the apparently irregular paths pursued by the planets when observed from the Earth. He knew of their stationary points, and also the backward loopings traced out by them on the surface of the sphere.

If observed from the Sun, all the planets would be seen to follow their true paths round that body; their motion would invariably lie in the same direction, and any variation in their speed as they approached perihelion or aphelion would be real. But the planets, when observed from the Earth, which is itself in motion, appear to move irregularly. Sometimes they remain stationary for a brief period, and, instead of progressing onward, affect a retrograde movement. This irregularity of motion is only apparent, and can be explained as a result of the combined motions of the Earth and planets, which are travelling together round the Sun with different velocities, and in orbits of unequal magnitude.

In his allusion to the Copernican system the 'planet' 'Earth' is described by Milton as seventh. This is not strictly accurate, as only five planets were known—viz. Mercury, Venus, Mars, Jupiter, and Saturn; but to make up the number Milton has included the Moon, which may be regarded as the Earth's planet.

The three motions ascribed to the Earth are—(1) The diurnal rotation on her axis; (2) her annual revolution round the Sun ; (3) Precession of the Equinoxes.

The rotation of the Earth on her axis may be likened to the spinning motion of a top, and is the cause of the alternation of day and night. This rotatory motion is sustained with such exact precision that, during the past 2,000 years, it has been impossible to detect the minutest difference in the time in which the Earth accomplishes a revolution on her axis, and therefore the length of the sidereal day, which is 3 minutes 56 seconds shorter than the mean solar day, is invariable. In this motion of the Earth we have a time-measuring unit which may be regarded as absolutely correct.

The Earth completes a revolution of her orbit in $365\frac{1}{4}$ days. In this period of time she accomplishes a journey of 580 millions of miles, travelling at the average rate of 66,000 miles an hour. The change of the seasons, and the lengthening and shortening of the day, are natural phenomena, which occur as a consequence of the Earth's annual revolution round the Sun. Precession is a retrograde or westerly motion of the equinoctial points, caused by the attraction of the Sun, Moon, and planets on the spheroidal figure of the Earth. By this movement the poles of the Earth are made to describe a circular path in that part of the heavens to which they point; so that, after the lapse of many years, the star which is known as the Pole Star will not occupy the position indicated by its name, but will be situated at a considerable distance from the pole. These motions, Milton says, unless attributed to the Earth, must be ascribed to several spheres crossing and thwarting each other obliquely; but the Earth, by rotating from west to east, will of herself fetch day, her other half, averted from the Sun's rays, being enveloped in night. Thus saving the Sun his labour, and the 'primum mobile,' 'that swift nocturnal and diurnal rhomb,' which carried all the lower spheres along with it, and brought about the change of day and night.

Milton's allusion to the occurrence of natural phenomena in the Moon similar to those which happen on the Earth is in keeping with the opinions entertained regarding our satellite, Galileo having imagined that he discovered with his telescope continents and seas on the lunar surface, which led to the belief that the Moon was the abode of intelligent life.

> . . . . and other suns, perhaps,
> With their attendant moons, thou wilt descry
> Communicating male and female light.—viii. 148–50.

Milton in these lines refers to Jupiter and Saturn, and their satellites, which had been recently discovered; those of the former by Galileo, and four of those of the latter by Cassini. The existence of male and female light was an idea entertained by the ancients, and which is mentioned by Pliny. The Sun was regarded as a masculine star, and the Moon as feminine; the light emanating from each being similarly distinguished, and possessing different properties.

Milton supposes that, as the Earth receives light from the stars, she returns light back to them. But in his time little was known about the stars, nor was it ascertained how distant they are.

The Angel, in bringing to a conclusion his conversation with Adam, deems it unadvisable to vouchsafe him a decisive reply to his inquiry regarding the motions of celestial bodies, and in the following lines gives a beautifully poetical summary of this elevated and philosophic discussion :—

But whether thus these things, or whether not,
Whether the Sun, predominant in Heaven,
Rise on the Earth, or Earth rise on the Sun ;
He from the east his flaming round begin,
Or she from west her silent course advance
With inoffensive pace that spinning sleeps
On her soft axle, whilst she paces even,
And bears thee soft with the smooth air along—
Solicit not they thoughts with matters hid.

                                    viii. 159–67.

In this scientific discourse between Adam and
Raphael, in which they discuss the structural
arrangement of the heavens and the motions of
celestial bodies, we are afforded an opportunity of
learning what exact and comprehensive knowledge
Milton possessed of both the Ptolemaic and Copernican theories.   The concise and accurate manner
in which he describes the doctrines belonging to
each system indicates that he must have devoted
considerable time and attention to making himself
master of the details associated with both theories,
which in his time were the cause of much controversy and discussion among philosophers and men
of science.

The Ptolemaic system, with its crystalline spheres
revolving round the Earth, the addition to those of
cycles and epicycles, and the heaping of them upon
each other, in order to account for phenomena
associated with the motions of celestial bodies, are
concisely and accurately described.

The unreasonableness of this theory, when compared with the Copernican, is clearly delineated by
Milton where Adam is made to express his views

with regard to motion in the heavens. His argument, declared in logical and persuasive language, demonstrates how contrary to reason it would be to imagine that the entire heavens should revolve round the Earth to bring about a result which could be more easily attained by imparting motion to the Earth herself. The inconceivable velocity with which it would be necessary for the celestial orbs to travel in order to accomplish their daily revolution is described by him as opposed to all reason, and entailing upon them a journey which it would be impossible for material bodies to perform. None the less accurate is Milton's description of the Copernican system. He describes the Sun as occupying that position in the system which his magnitude and supreme importance claim as his sole right, having the planets with their satellites,

That from his lordly eye keep distance due.—iii. 578,

circling in majestic orbits around him, acknowledging his controlling power, and bending to his firm but gentle sway. Their positions, their paths, and their motions, real and apparent, are described in flowing and harmonious verse.

CHAPTER IV

MILTON AND GALILEO

AFTER the death of his mother, which occurred in 1637, Milton expressed a desire to visit the Continent, where there were many places of interest which he often longed to see. Having obtained the consent of his kind and indulgent father, he set out on his travels in April 1638, accompanied by a single man-servant, and arrived in Paris, where he only stayed a few days. During his residence in the French capital he was introduced by Lord Scudamore, the English Ambassador at the Court of Versailles, to Hugo Grotius, one of the most distinguished scholars and philosophic thinkers of his age. From Paris Milton journeyed to Nice, where he first beheld the beauty of Italian scenery and the classic shores of the Mediterranean Sea. From Nice he sailed to Genoa and Leghorn, and after a short stay at those places continued his journey to Florence, one of the most interesting and picturesque of Italian cities. Situated in the Valley of the Arno, and encircled by sloping hills covered with luxuriant vegetation, the sides of which were studded with residences half-hidden among the foliage of gardens and vineyards, Florence, besides being famed for

I

its natural beauty, was at that time the centre of
Italian culture and learning, and the abode of men
eminent in literature and science. Here Milton
remained for a period of two months, and enjoyed
the friendship and hospitality of its most noted
citizens, many of whom delighted to honour their
English visitor. He was warmly welcomed by the
members of the various literary academies, who
admired his compositions and conversation ; the
flattering encomiums bestowed upon him by those
learned societies having been amply repaid by Milton
in choice and elegant Latin verse.

Among those who resided in the vicinity of
Florence was the illustrious Galileo, who in his
sorrow-stricken old age was held a prisoner of the
Inquisition for having upheld and taught scientific
doctrines which were declared to be heretical.
After his abjuration he was committed to prison,
but on the intervention of influential friends was
released after a few days' incarceration, and permitted to return to his home at Arcetri. He was,
however, kept under strict surveillance, and forbidden to leave his house or receive any of his
intimate friends without having first obtained the
sanction of the ecclesiastical authorities. After
several years of close confinement at Arcetri, during
which time he suffered much from rheumatism and
continued ill-health, aggravated by grief and mental
depression consequent upon the death of his
favourite daughter, Galileo applied for permission
to go to Florence in order to place himself under

medical treatment. This request was granted by
the Pope subject to certain conditions, which would
be communicated to him when he presented himself at the office of the Inquisition at Florence.
These were more severe than he anticipated. He
was forbidden to leave his house or receive any of
his friends there, and those injunctions were so
strictly adhered to that during Passion Week he
had to obtain a special order so that he might be
able to attend mass. At the expiration of a few
months Galileo was ordered to return to Arcetri,
which he never left again.

An affliction, perhaps the most deplorable that
can happen to any human being, was added
to the burden of Galileo's misfortunes and woes.
A disorder which had some years previously
injured the sight of his right eye returned in
1636. In the following year the left eye became
similarly affected, with the result that in a few
months Galileo became totally blind. His friends
at first hoped that the disease was cataract, and
that some relief might be afforded by means of an
operation ; but it was discovered to be an opacity
of the cornea, which at his age was considered
unamenable to treatment. This sudden and unexpected calamity was to Galileo a most deplorable
occurrence, for it necessitated the relinquishment of
his favourite pursuit, which he followed with such
intense interest and delight. His friend Castelli
writes : ' The noblest eye is darkened which Nature
ever made ; an eye so privileged, and gifted with

such rare qualities that it may with truth be said
to have seen more than all of those eyes who are
gone, and to have opened the eyes of all who are
to come.'   Galileo endured his affliction with patient resignation and fortitude, and in the following extract from a letter by him he acknowledges
the chastening hand of a Divine Providence:
'Alas! your dear friend and servant Galileo has
become totally blind, so that this heaven, this earth,
this universe, which with wonderful observations
I had enlarged a hundred and a thousand times
beyond the belief of bygone ages, henceforward
for me is shrunk into the narrow space which I
myself fill in it.   So it pleases God; it shall then
please me also.'   The rigorous curtailment of his
liberty which prompted Galileo to head his letters,
'From my prison at Arcetri,' was relaxed when
total blindness had supervened upon the infirmities
of age.   Permission was given him to receive his
friends, and he was allowed to have free intercourse
with his neighbours.

Milton, during his stay at Florence, visited
Galileo at Arcetri.   We are ignorant of the details
of this eventful and interesting interview between
the aged and blind astronomer and the young English poet, who afterwards immortalised his name in
heroic verse, and who in his declining years suffered
from an affliction similar to that which befel Galileo,
and to which he alludes so pathetically in the
following lines :—

Thee I revisit safe,
And feel thy sovran vital lamp; but thou

Revisitest not these eyes, that roll in vain
To find thy piercing ray, and find no dawn ;
So thick a drop serene hath quenched their orbs,
Or dim suffusion veiled.—iii. 21-26.

We can imagine that Galileo's astronomical
views, which at that time were the subject of much
discussion among scientific men and professors of
religion, and on account of which he suffered
persecution, were eagerly discussed. It is also
probable that the information communicated by
Galileo, or by some of his followers, may have persuaded Milton to entertain a more favourable
opinion of the Copernican theory. The interesting
discoveries made by Galileo with his telescope
without doubt formed a pleasant subject of conversation, and Milton enjoyed the privilege of listening
to a detailed description of these from the lips of
the aged astronomer. The telescope, its principle,
its mechanism, and the method of observing, were
most probably explained to him ; and we can believe
that an opportunity was afforded him of examining
those in Galileo's observatory, and of perhaps testing
their magnifying power upon some celestial object
favourably situated for observation. Though Milton
has not favoured us with any details of his visit to
Galileo, yet it was one which made a lasting impression upon his mind, and was never afterwards forgotten by him. ' There it was,' he writes, ' I found
and visited the famous Galileo, grown old, a prisoner
of the Inquisition for thinking in astronomy otherwise than the Franciscan and Dominican licensers
thought.' In years long after, when Milton, himself

feeble and blind, sat down to compose his ' Paradise
Lost,' the remembrance of the Tuscan artist and
his telescope was still fresh in his memory.

By the invention of the telescope and its application to astronomical research, a vast amount
of information and additional detail have been
learned regarding the bodies which enter into the
formation of the solar system ; and by its aid
many new ones were also discovered.   On sweeping
the heavens with the instrument, the illimitable
extent of the sidereal universe became apparent,
and numberless objects of interest were brought
within the range of vision the existence of which
had not been previously imagined.

The Galilean telescope was invented in 1609.
But the magnifying power of certain lenses, and
their combination in producing singular visual
effects, are alluded to in the writings of several early
authors.   The value of single lenses as an aid to
sight had been long known, and spectacles were in
common use in the fourteenth century.   Several
mathematicians have described the wonderful optical
results obtained from glasses concave and convex, of
parabolic and circular forms, and from ' perspective
glasses,' in which were embodied the principle of
the telescope.   It is asserted that our countryman,
Roger Bacon (1214), had some notion of the properties of the telescope ; but among those familiar
with the combination of lenses the two men who
made the nearest approach to the invention of the
instrument were Baptista Porta and Gerolamo

Fracastro. The latter, who died in 1553, writes as follows : ' For which reason those things which are seen at the bottom of water appear greater than those which are at the top; and if anyone look through two eye-glasses, one placed upon the other, he will see everything much larger and nearer.' It is doubtful if Fracastro had any notion of constructing a mechanism which might answer the purpose of a telescopic tube. Baptista Porta (1611) is more explicit in what he describes. He writes : ' Concave lenses show distant objects most clearly, convex those which are nearer; whence they may be used to assist the sight. With a concave glass distant objects will be seen, small, but distinct; with a convex one, those near at hand, larger, but confused; if you know *rightly* how to combine one of each sort, you will see both far and near objects larger and clearer.' He then goes on to say : ' I shall now endeavour to show in what manner we may continue to recognise our friends at the distance of several miles, and how those of weak sight may read the most minute letters from a distance. It is an invention of great utility, and grounded on optical principles ; nor is at all difficult of execution ; but it must be so divulged as not to be understood by the vulgar, and yet be clear to the sharp-sighted.' After this, he proceeds to describe a mechanism the details of which are confusing and unintelligible, nor did it appear to bear any resemblance to a telescopic tube.

In a work published by Thomas Digges in 1591,

he makes the following allusion to his father's
experiments with the lenses : ' My father, by his
continuall painfull practices, assisted with demon-
strations mathematicall, was able, and sundry
times hath by proportionall glasses, duely situate
in convenient angles, not only discouered things
farre off, read letters, numbered peeces of money
with the verye coyne and superscription thereof
cast by some of his freends of purpose, upon downes
in open fields ; but also suen miles off, declared
what hath beene doone at that instant in priuate
places.'   It must be admitted that if Leonard
Digges had not constructed a telescope, he knew
how to combine lenses by the aid of which a visual
effect was created similar to that produced by the
use of the instrument.

The inventor of the telescope was a Dutchman
named Hans Lippershey, who carried on the business of a spectacle-maker in the town of Middelburg.
His discovery was purely accidental. It is said
that the instrument—which was directed towards
a weather-cock on a church spire, of which it gave
a large and inverted image—was for some time
exhibited in his shop as a curiosity before its importance was recognised. The Marquis Spinola,
happening to see this philosophical toy, purchased
it, and presented it to Prince Maurice of Nassau,
who imagined it might be of service for the purpose of military reconnoitring. The value of the
invention was, however, soon realised, and in the
following year telescopes were sold in Paris. In

1609, Galileo, when on a visit to a friend at Venice,
received intelligence of the invention of an instrument by a Dutch optician which possessed the
power of causing distant objects to appear much
nearer than when observed by ordinary vision.
The accuracy of this information was confirmed
by letters which he received from Paris ; and this
general report, Galileo asserted, was all he knew
of the subject.   Fuccarius, in a disparaging letter,
says that one of the Dutch telescopes had been
brought to Venice, and that he himself had seen it.
This statement is not incompatible with Galileo's
affirmation that he had not seen the original instrument, and knew no more about it than what had
been communicated to him in the letters from the
French capital.   It was insinuated by Fuccarius
that Galileo had seen the telescope at Venice, but,
as he denied this, we should not hesitate to believe
in his veracity.

Immediately after his return to Padua, Galileo
began to think how he might be able to contrive
an instrument with properties similar to the one of
which he had been informed ; and in the following
words describes the process of reasoning by which
he arrived at a successful result : ' I argued in
the following manner.   The contrivance consists
either of one glass or of more—one is not sufficient,
since it must be either convex, concave, or plane.
The last does not produce any sensible alteration
in objects ; the concave diminishes them.   It is
true that the convex magnifies, but it renders them

confused and indistinct; consequently, one glass
is insufficient to produce the desired effect.   Proceeding to consider two glasses, and bearing in
mind that the plane causes no change, I determined
that the instrument could not consist of the combination of a plane glass with either of the other
two.   I therefore applied myself to make experiments on combinations of the two other kinds, and
thus obtained that of which I was in search.'
Galileo's telescope consisted of two lenses—one
plano-convex, the other plano-concave, the latter
being held next the eye.   These he fixed in a piece
of organ pipe, which served the purpose of a tube,
the glasses being distant from each other by the
difference of their focal lengths.   An exactly
similar principle is adopted in the construction of an
opera-glass, which can be accurately described as a
double Galilean telescope.   Galileo must be regarded
as the inventor of this kind of telescope, which in
one respect differed very materially from the one
constructed by the Dutch optician.   If what has
been said with regard to the *inverted* weathercock
be true, then Lippershey's telescope was made with
two convex lenses, distant from each other by the
sum of their focal lengths, and all objects observed
with it were seen inverted.   Refracting astronomical telescopes are now constructed on this principle,
it having been discovered that for observational
purposes they possess several advantages over the
Galilean instrument.   When Galileo had completed
his first telescope he returned with it to Venice,

where he exhibited it to his friends.   The sensation
created by this small instrument, which magnified
only three times, was most extraordinary, and almost amounted to a frenzy.   Crowds of the principal citizens of Venice flocked to Galileo's house
in order that they might see the magical tube
about which such wonderful reports were circulated ;
and for upwards of a month he was daily occupied
in describing his invention to attentive audiences.
At the expiration of this time the Doge of Venice,
Leonardo Deodati, hinted that the Senate would
not be averse to receive the telescope as a gift.
Galileo readily acquiesced with this desire, and,
as an acknowledgment of his merits, a decree was
issued confirming his appointment as professor at
Padua for life, and increasing his salary from 500
to 1,000 florins.   The public excitement created
by the telescope showed no signs of abatement.
Sirturi mentions that, having succeeded in constructing an instrument, he ascended the tower of
St. Mark's at Venice, hoping to be able to use it
there without interruption.   He was, however,
detected by a few individuals, and soon surrounded
by a crowd, which took possession of his telescope,
and detained him for several hours until their curiosity was satisfied.   Eager inquiries having been
made as to where he lodged, Sirturi, fearing a repe-
tition of his experience in the church tower, decided
to quit Venice early next morning, and betake
himself to a quieter and less frequented neighbourhood.

The instrument was at first called Galileo's tube; the double eye-glass; the perspective; the trunk; the cylinder. The appellation *telescope* was given it by Demisiano.

Galileo next directed his attention to the construction of telescopes, and applied his mechanical skill in making instruments of a larger size, one of which magnified *eight* times. 'And at length,' he writes, 'sparing neither labour nor expense, he completed an instrument that was capable of magnifying more than *thirty* times.'

Galileo now commenced an exploration of the celestial regions with his telescope, and on carefully examining some of the heavenly bodies, made many wonderful discoveries which added greatly to the fame and lustre of his name.

The first celestial object to which Galileo directed his telescope was the Moon. He was deeply interested to find how much her surface resembled that of the Earth, and was able to perceive lofty mountain ranges, the illumined peaks of which reflected the sunlight, whilst their bases and sides were still enveloped in dark shadow; great plains which he imagined were seas, valleys, elevated ridges, depressions, and inequalities similar to what are found on our globe. Galileo believed the Moon to be a habitable world, and concluded that the dark and luminous portions of her surface were land and water, which reflected with unequal intensity the light of the Sun. The followers of Aristotle received the announcement of these dis-

coveries with much displeasure. They maintained
that the Moon was perfectly spherical and smooth
—a vast mirror, the dark portions of which were
the reflection of our terrestrial mountains and
forests—and accused Galileo ' of taking a delight in
distorting and ruining the fairest works of Nature.'
He appealed to the unequal condition of the surface
of our globe, but this was of no avail in altering
their preconceived notions of the lunar surface.

Perhaps the most important discovery made by
Galileo with the telescope was that of the four
moons of Jupiter. On the night of January 7, 1610,
when engaged in observing the planet, his attention
was attracted by three small stars which appeared
brighter than those in their immediate neighbourhood. They were all in a straight line and parallel
with the ecliptic; two of them were situated to
the east, and one to the west of Jupiter. On the
following night he was surprised to find all three to
the west of the planet, and nearer to each other.
This caused him considerable perplexity, and he
was at a loss to understand how Jupiter could be
east of the three stars, when on the preceding night
he was observed to the west of two of them. Galileo
was unable to reconcile the altered positions of
those bodies with the apparent motion of Jupiter
among the fixed stars as indicated by the astronomical tables. The next opportunity he had of
observing them was on the 10th, when two stars
only were visible, and they were to the east of the
planet. As it was impossible for Jupiter to move

from west to east on January 8 and from east to west on the 10th, he concluded that it was the motion of the stars and not that of Jupiter which accounted for the observed phenomena. Galileo watched the stars attentively on successive evenings and discovered a fourth, and on observing how they changed their positions relatively to each other he soon arrived at the conclusion that the stars were four moons which revolved round Jupiter after the manner in which the Moon revolves round the Earth. Having assured himself that the four new stars were four moons that with periodical regularity circled round the great planet, Galileo named them the Medicean Stars in honour of his patron, Cosmo de' Medici, Grand Duke of Tuscany. He also published an essay entitled 'Nuncius Sidereus,' or the 'Sidereal Messenger,' which contained an account of this important discovery.

The announcement of Galileo's discovery of the four satellites of Jupiter created a profound sensation, and its significance became at once apparent. Aristotelians and Ptolemaists received the information with much disfavour and incredulity, and many persons positively refused to believe Galileo, whom they accused of inventing fables. On the other hand, the upholders of the Copernican theory hailed it with satisfaction, as it declared that Jupiter with his four moons constituted a system of greater magnitude and importance than that of our globe with her single satellite, and that consequently the Earth could not be regarded as the centre of the universe.

When Kepler heard of this remarkable discovery, he wrote to Galileo and expressed himself in the following characteristic manner : ' I was sitting idle at home thinking of you, most excellent Galileo, and your letters, when the news was brought me of the discovery of four planets by the help of the double eye-glass. Wachenfels stopped his carriage at my door to tell me, when such a fit of wonder seized me at a report which seemed so very absurd, and I was thrown into such agitation at seeing an old dispute between us decided in this way, that between his joy, my colouring, and the laughter of both, confounded as we were by such a novelty, we were hardly capable, he of speaking, or I of listening. . . . I am so far from disbelieving in the existence of the four circumjovial planets, that I long for a telescope to anticipate you, if possible, in discovering two round Mars (as the proportion seems to me to require), six or eight round Saturn, and perhaps one each round Mercury and Venus.' The intelligence of Galileo's discoveries was received by his opponents in a spirit entirely different from that manifested by Kepler. The principal professor of philosophy at Padua, when requested to look at the Moon and planets through Galileo's glass, persistently declined, and did his utmost to persuade the Grand Duke that the four satellites of Jupiter could not possibly exist. Francesco Sizzi, a Florentine astronomer, argued that, as there are seven

apertures in the head, seven known metals, and
seven days in the week, so there could only be seven
planets. To these absurd remarks Galileo replied
by saying that, 'whatever their force might be as
a reason for believing beforehand that no more than
seven planets would be discovered, they hardly
seemed of sufficient weight to destroy the new
ones when actually seen.' Another individual,
named Christmann, writes : ' We are not to think
that Jupiter has four satellites given him by Nature
in order, by revolving round him, to immortalize the
name of the Medici, who first had notice of the
observation. These are the dreams of idle men,
who love ludicrous ideas better than our laborious
and industrious correction of the heavens. Nature
abhors so horrible a chaos, and to the truly wise
such vanity is detestable.' Martin Horky, a *protégé*
of Kepler's, issued a pamphlet in which he made a
violent attack on Galileo. He says : ' I will never
concede his four new planets to that Italian from
Padua though I die for it.' He then asks the
following questions, and replies to them himself :
(1) Whether they exist ? (2) What they are ?
(3) What they are like ? (4) Why they are ?
' The first question is soon disposed of by Horky's
declaring positively that he has examined the
heavens with Galileo's own glass, and that no such
thing as a satellite about Jupiter exists. To the
second, he declared solemnly that he does not more
surely know that he has a soul in his body than
that reflected rays are the sole cause of Galileo's

erroneous observations. In regard to the third question, he says that these planets are like the smallest fly compared to an elephant; and, finally, concludes on the fourth, that the only use of them is to gratify Galileo's "thirst of gold," and to afford himself a subject of discussion.'[1] Galileo did not condescend to take any notice of this scurrilous production; but Horky, who imagined that he had done something clever, sent a copy of his pamphlet to Kepler. In a few days after he called to see him, and was received with such a storm of indignation that he begged for mercy and implored his forgiveness. Kepler forgave him, but insisted on his making amends. He writes: 'I have taken him again into favour upon this preliminary condition, to which he has agreed—that I am to show him Jupiter's satellites, *and he is to see them*, and own that they are there.'

The evidence in support of the existence of Jupiter's satellites became so conclusive that the opponents of Galileo were compelled to renounce their disbelief in those bodies, whether real or pretended. The Grand Duke, preferring to trust to his eyes rather than believe in the arguments of the professor at Padua, observed the satellites on several occasions, along with Galileo, at Pisa, and on his departure bestowed upon him a gift of one thousand florins. Several of Galileo's enemies, as a result of their observations, now arrived at the conclusion that his discovery was

[1] *Life of Galileo* (Library of Useful Knowledge).

incomplete, and that Jupiter had more than four satellites in attendance upon him. Scheiner counted five, Rheita nine, and other observers increased the number to twelve. But it was found to be quite as hazardous to exceed the number stated by Galileo as it was to deny the existence of any; for, when Jupiter had traversed a short distance of his path among the fixed stars, the only bodies that accompanied him were his four original attendants, which continued to revolve round him with unerring regularity in every part of his orbit.

Galileo did not afford his opponents much time to oppose or controvert with argument the discoveries made by him with the telescope before his announcement of a new one attracted public attention from those already known. He, however, exercised greater caution in disclosing the results of his observations, as other persons laid claim to having made similar discoveries prior to the time at which his were announced. He therefore adopted a method in common use among astronomers in those days, by which the letters in a sentence announcing a discovery were transposed so as to form an anagram.

Galileo announced his next discovery in this manner, and which read as follows :—

Smaismrmilme poeta leumi bvne nugttaviras.

This, when deciphered, formed the sentence :—

Altissimum planetam tergeminum observavi.

I have observed that the remotest planet is triple.

Galileo perceived that Saturn presented a triform appearance, and that, instead of one body, there were three, all in a straight line, and apparently in contact with each other, the middle one being larger than the two lateral ones. In a letter to Kepler he remarked: ' Now I have discovered a Court for Jupiter, and two servants for this old man, who aid his steps and never quit his side.' Kepler, who excelled as an imaginative writer, replied: ' I will not make an old man of Saturn, nor slaves of his attendant globes; but rather let this tricorporate form be Geryon—so shall Galileo be Hercules, and the telescope his club, armed with which he has conquered that distant planet, and dragged him from the remotest depths of Nature, and exposed him to the view of all.' Continuing his observations, Galileo perceived that the two lateral objects gradually decreased in size, and at the expiration of two years entirely disappeared, leaving the central globe visible only. He was unable to assign any reason for this peculiar occurrence, which caused him much perplexity, and he expresses himself thus : ' What is to be said concerning so strange a metamorphosis ? Are the two lesser stars consumed after the manner of the solar spots ? Have they vanished and suddenly fled ? Has Saturn, perhaps, devoured his own children ? Or were the appearances, indeed, illusion or fraud, with which the glasses have so long deceived me, as well as many others to whom I have shown them ? Now, perhaps, is the time to revive the well-nigh withered hopes of

those who, guided by more profound contempla-
tions, have discovered the fallacy of the new observations, and demonstrated the utter impossibility
of their existence.   I do not know what to say in a
case so surprising, so unlooked-for, and so novel.
The shortness of the time, the unexpected nature
of the event, the weakness of my understanding,
and the fear of being mistaken, have greatly confounded me.'   After a certain interval those bodies
reappeared; but Galileo's glass was not sufficiently powerful to enable him to ascertain their
nature nor solve the mystery, which for upwards of
half a century perplexed the ablest astronomers.

The elucidation of this inexplicable phenomenon
was reserved for Christian Huygens, who, with an
improved telescope of his own construction, was able
to declare that Saturn's appendages were portions
of a ring which surrounds the planet, and is everywhere distinct from its surface.

Galileo next directed his attention to the planet
Venus, and as a result of his observations was led
to communicate to the public another anagram :—

> Haec immatura a me jam frustra leguntur oy.

This, when rendered correctly, reads :—

> Cynthiae figuras aemulatur mater amorum.
> Venus rivals the appearances of the Moon.

The phases of Venus were one of the most interesting of Galileo's discoveries with the telescope.
When observed near inferior conjunction the planet
presents the appearance of a slender crescent, re-

sembling the Moon when a few days old. Travelling from this point to superior conjunction, the illumined portion of her disc gradually increases, until it becomes circular, like the full Moon. This changing appearance of Venus afforded Galileo irresistible proof that the planet is an opaque body, which derives its light from the Sun, and that it circles round the orb—convincing evidence of the accuracy and truthfulness of the Copernican theory.

It was in this manner that Galileo announced his discovery of the phases of Venus, the peerless planet of our morning and evening skies, whose slender crescent forms such a beautiful object in the telescope, and who, as she traverses her orbit, exhibits all the varied changes of form presented by the Moon in her monthly journey round the Earth. These varying aspects of Venus were not unknown to Milton; and, indeed, he may have been informed of them by Galileo in his conversation with him at Arcetri; nor has he failed to introduce an allusion to this beautiful phenomenon in his poem. In his description of the Creation, after the Sun was formed, he adds :—

> Hither, as to their fountain, other stars
> Repairing, in their golden urns draw light,
> And hence the morning planet gilds her horns.
>
> vii. 364–66.

Galileo also discovered that the planet Mars does not always present the appearance of a circular disc. When near opposition the full disc of the planet is visible, but at all other times it is gibbous, and approaches nearest to that of a half-moon when at the quadratures.

In the year 1610, on directing his telescope to the Sun, Galileo detected dark spots on the solar disc. Similar spots, sufficiently large to be distinguished by the naked eye, had been observed from time to time for centuries prior to the invention of the telescope, but nothing was known of their nature. In 1609 Kepler observed a spot on the Sun, which he thought was the planet Mercury in conjunction with the orb; the short time during which it was visible, in consequenc of clouds having obscured the face of the luminary, prevented him from being able to determine the accuracy of his surmise, but since then it has been ascertained that no transit of Mercury took place at that time, and Kepler afterwards acknowledged that he had arrived at an erroneous conclusion. Galileo was much puzzled in trying to find out the true nature of the spots. At first he was led to imagine that planets like Mercury and Venus revolved round the Sun at a short distance from the orb, and that their dark bodies, travelling across the solar disc, gave rise to the phenomenon of the spots. After further observation, he ascertained that the spots were in actual contact with the Sun; that they were irregular in shape and size, and continued to appear and disappear. Sometimes a large spot would break up into several smaller ones, and at other times three or four small spots would unite to form a large one. They all had a common motion, and appeared to rotate with the Sun, from which Galileo concluded that the orb rotated on his axis in about twenty-eight days. Galileo believed that the spots were clouds floating in the solar atmosphere, and that they intercepted a portion of the light of the Sun.

The Milky Way, that wondrous zone of light which encircles the heavens, remained for many ages a source of perplexity to ancient astronomers and philosophers, who, in their endeavours to ascertain its nature, had arrived at various absurd and erroneous conclusions. On directing his telescope to this luminous tract, Galileo discovered, to his inexpressible admiration, that it consists of a vast multitude of stars, too minute to be visible to the naked eye. He also discerned that its milky luminosity is created by the blended light of myriads of stars, so remote as to be incapable of definition by his telescope. In his ' Nuncius Sidereus ' he gives an account of his observations of the Galaxy and expresses his satisfaction that he has been enabled to terminate an ancient controversy by demonstrating to the senses the stellar structure of the Milky Way. When engaged in exploring the celestial regions with his telescope, Galileo observed a marked difference in the appearance of the fixed stars, as compared with that of the planets. Each of the latter showed a rounded disc resembling that of a small moon, but the stars exhibited no disc, and shone as vivid sparkling points of light; all of them, whether of large or small magnitude, presenting the same appearance in the telescope. This led him to conclude that the fixed stars were not illumined by the Sun, because their brilliancy in all their changes of position remained unaltered. But, in the case of the planets, he found that their lustre varied according to their distance from the Sun ; consequently, he believed they were opaque bodies which reflected the solar rays. On directing his telescope to the Pleiades, which, to the naked eye, appear as a group of seven stars, he succeeded in counting forty lucid points. The nebula Praesepe in Cancer, he was also able to resolve into a cluster of stars. Galileo made many other observations of the heavenly bodies with his telescope, all of which he describes as having afforded him 'incredible delight.'

Shortly before the failure of his eyesight, Galileo discovered the Moon's diurnal libration, a variation in the visible edges of the Moon caused by its oscillatory motion, and the diurnal rotation of the Earth on her axis.

Though Milton has not favoured us with any interesting details of his interview with Galileo, nor expressed his opinions with regard to the controversies which at that time agitated both the religious and scientific worlds of thought, and which eventually culminated in a storm of rancour and hatred that burst over the devoted head of the aged astronomer, and brought him to his knees, yet he informs us that he 'found and visited' Galileo, whom he describes as 'grown old,' and cynically remarks that he ' was held a prisoner of the Inquisition for thinking in astronomy otherwise than the Franciscan and Dominican licensers thought.' Milton does not allude to his blindness, and yet it would be natural to imagine that, had his host suffered from this affliction at the time of his visit, he would have referred to it. We learn that Milton arrived in Italy in the spring of 1638. In 1637, the affection which, in the preceding year, deprived Galileo of the use of his right eye, attacked the left also, which began to grow dim, and in the course of a few months became sightless; so that, although Milton has not alluded to this calamity, Galileo had become totally blind at the time of his visit.

How much Milton was impressed with the fame of Galileo and his telescope becomes apparent on referring to his ' Paradise Lost.' In it he alludes to the instrument upon three different occasions, twice when in the hands of Galileo; and the remembrance of the same artist was doubtless in his mind when he mentions the 'glazed optic tube' in another part of his poem. The interval that elapsed from the date of Milton's visit to Galileo in 1638, to the publication of ' Paradise Lost ' in 1667, included a period of about thirty years, yet this length of time did not erase from Milton's memory his recollection of Galileo and of his pleasant sojourn at Florence.

The first allusion in the poem to the Italian astronomer is in the lines in which Milton describes the shield carried by Satan :—

> The broad circumference
> Hung on his shoulders like the Moon, whose orb
> Through optic glass the Tuscan artist views
> At evening, from the top of Fesolé,
> Or in Valdarno, to descry new lands,
> Rivers, or mountains, in her spotty globe.—i. 286–91.

Galileo is described as having observed the Moon from the heights of Fesolé, which formed part of the suburbs of Florence, or from Valdarno, the valley of the Arno, in which the city is situated. The belief that Galileo had discovered continents and seas on the Moon justified Milton in imagining the existence of rivers and mountains on the lunar surface. The expression 'spotty globe' is more descriptive of the appearance of our satellite when observed with the telescope, than when seen with the naked eye. Galileo's attention was attracted by the freckled aspect of the Moon—a visual effect created by the number of extinct volcanoes scattered over the surface of the orb.

In his next allusion to the telescope Milton associates Galileo's name with the instrument :—

> As when by night the glass
> Of Galileo, less assured, observes
> Imagined lands and regions in the Moon.—v. 261–63.

In these lines Milton describes with accuracy the extent of Galileo's knowledge of our satellite. The conclusions which the Italian astronomer arrived at with regard to its habitability were not supported by telescopic evidence sufficient to justify such a belief. Galileo writes : ' Had its surface been absolutely smooth it would have been but a vast, unblessed desert, void of animals, of plants, of cities and men ; the abode of silence and inaction—senseless, lifeless, soulless, and stripped of all those ornaments which now render it so variable and so beautiful : '—

> There lands the Fiend, a spot like which perhaps
> Astronomer in the Sun's lucent orb
> Through his glazed optic tube yet never saw.
>
> iii. 588-90.

Milton may have remembered that Galileo was the first astronomer who directed a telescope to the Sun ; and that he discovered the dark spots frequently seen on the solar disc.

Anyone who has read a history of the life of Galileo, and contemplated the career of this remarkable man, his ardent struggles in the cause of freedom and philosophic truth, his victories and reverses, his brilliant astronomical discoveries, and his investigation of the laws of motion, and other natural phenomena, will arrive at the conclusion that he merited the distinction conferred upon him by our great English poet, when he included him among the renowned few whose names are found in the pages of 'Paradise Lost.'

CHAPTER V

THE SEASONS

THE great path of the Sun among the constellations as seen from the Earth is called the Ecliptic. It is divided into 360°, and again into twelve equal parts of 30°, called Signs. As one half of the ecliptic is north, and the other half south, of the equator, the line of intersection of their planes is at two points which are known as the equinoctial points, because, when the Sun on his upward and downward journey arrives at either of them the days and nights are of equal length all over the world. The equinoctial points are not stationary, but have a westerly motion of 50″ annually along the ecliptic ; at this rate they will require a period of 25,868 years to complete an entire circuit of the heavens.

Milton alludes to the ecliptic when he mentions the arrival of Satan upon the Earth :—

> Down from the ecliptic, sped with hoped success,
> Throws his steep flight in many an airy wheel,
> Nor staid till on Niphates top he lights.—iii. 740-42.

Extending for 9° on each side of the ecliptic is a zone or belt called the Zodiac, the mesial line of which is occupied by the Sun, and within this space the principal planets perform their annual revolutions. It was for long believed that the paths of all the planets lay within the zodiac, but on the discovery of the minor planets, Ceres, Pallas, and Juno, it was ascertained that they travelled beyond this zone. The stars situated within the zodiac are divided into twelve groups or constellations, which correspond with the twelve signs, and each is named after an animal or some figure which it is supposed to resemble. The zodiac is of great antiquity; the ancient Egyptians and Hindoos made use of it, and there are allusions to it in the earliest astronomical records. The twelve constellations of the zodiac bear the following names :—

Aries . . .	the Ram	Scorpio . .	the Scorpion
Taurus . .	the Bull	Sagittarius .	the Archer
Gemini . .	the Twins	Capricornus .	the Goat
Cancer . .	the Crab	Aquarius . .	the Water-
Leo . . .	the Lion		bearer
Virgo . . .	the Virgin	Pisces . .	the Fishes
Libra . . .	the Balance		

In close association with the Sun's annual journey are the seasons, upon the regular sequence of which mankind depend for the various products of the soil essential for the maintenance and enjoyment of life. The revolution of the Earth in her orbit, and the inclination of her axis to her annual path, causing the plane of the equator to be inclined $23\frac{1}{2}°$ to that of the ecliptic, are the reasons which account for the succession of the seasons—Spring, Summer, Autumn, and Winter. Owing to the position of the Earth's axis with regard to her orbit, the Sun appears to travel $23\frac{1}{2}°$ north and $23\frac{1}{2}°$ south of the equator. When, on June 21, the orb attains his highest northern altitude, we have the summer solstice and the longest days; when, by retracing his steps, he declines $23\frac{1}{2}°$ below the equator, at which point he arrives on December 21, we have the winter solstice and the shortest days. Intermediate between those two seasons are spring and autumn. When the Sun, on his journey northward, reaches the equator, we have the vernal equinox, and at this period of the year the days and nights are of equal length all over the globe. In a similar manner, when, on his return journey, the Sun is again on the equator, the autumnal equinox occurs. In summer the North Pole is inclined towards the Sun, consequently his rays fall more direct and impart much more heat to the northern hemisphere than in winter, when the Pole is turned away from the Sun. This difference in the incidence of the solar rays upon the surface of the globe, along with the increased length of the day, mainly accounts for the high temperature of summer as compared with that of winter.

Astronomically, the seasons commence at the periods of the equinoxes and solstices. Spring begins on March 21, the time of the vernal equinox ; summer on June 21, at the summer solstice ; autumn on September 22, at the autumnal equinox ; and winter on December 21, at the winter solstice. This conventional division of the year is not equally applicable to all parts of the globe. In the arctic and antarctic regions spring and autumn are very brief, the summer is short and the winter of long duration. In the tropics, owing to the comparatively slight difference in the obliquity of the Sun's rays, one season is, as regards temperature, not much different from the other; but in the temperate regions of the Earth the vicissitudes of the seasons are more perceptible and can be best distinguished by the growth of vegetation, and the changes observable in the foliage of shrubs and trees. In spring there is the budding, in summer the blossom, in autumn the fruit-bearing, and in winter the leafless condition of deciduous trees, and the repose of vegetable life.

The legendary belief that before the Fall there reigned on the Earth a perpetual spring, is introduced by Milton in his poem when he describes the pleasant surroundings associated with the happy conditions of life that existed in Paradise :—

> Thus was this place,
> A happy rural seat of various view :
> Groves whose rich trees wept odorous gums and balm ;
> Others whose fruit, burnished with golden rind,
> Hung amiable—Hesperian fables true,
> If true here only—and of delicious taste.
> Betwixt them lawns, or level downs, and flocks
> Grazing the tender herb, were interposed,
> Or palmy hillock ; or the flowery lap
> Of some irriguous valley spread her store,
> Flowers of all hue, and without thorn the rose.
> Another side, umbrageous grots and caves
> Of cool recess, o'er which the mantling vine
> Lays forth her purple grape, and gently creeps
> Luxuriant ; meanwhile murmuring waters fall
> Down the slope hill dispersed, or in a lake

That to the fringèd bank with myrtle crowned
Her crystal mirror holds, unite their streams.
The birds their quire apply ; airs, vernal airs,
Breathing the smell of field and grove, attune
The trembling leaves, while universal Pan,
Knit with the Graces and the Hours in dance,
Led on the eternal Spring.—iv. 246-68.

In sad contrast with this charming sylvan scene,
we turn to the unhappy consequences which ensued
as a result of the first act of transgression.   Milton
describes a change of climate characterised by
extremes of heat and cold which succeeded the
perpetual spring.   The Sun was made to shine so
that the Earth should be exposed to torrid heat
and icy cold unpleasant to endure.   The pale
Moon and the planets were given power to combine
with noxious effect, and the fixed stars to shed
their malignant influences :—

                                        The Sun
Had first his precept so to move, so shine,
As might affect the Earth with cold and heat
Scarce tolerable, and from the north to call
Decrepit winter, from the south to bring
Solstitial summer's heat.   To the blanc Moon
Her office they prescribed ; to the other five
Their planetary motions and aspects,
In sextile, square, and trine, and opposite,
Of noxious efficacy, and when to join
In synod unbenign ; and taught the fixed
Their influence malignant when to shower—
Which of them rising with the Sun or falling,
Should prove tempestuous.   To the winds they set
Their corners, when with bluster to confound
Sea, air, and shore ; the thunder when to roll
With terror through the dark aerial hall.—x. 651-67.

We are here afforded an opportunity of learning that Milton possessed some knowledge of astrology, to which he makes allusion in other parts of his poem besides. In his time, astrology was believed in by many persons, and there were few learned men but who knew something of that occult science. Milton may be included among those who devoted some attention to astrology. Of this there is ample evidence, by the manner in which he expresses himself in words and phrases in common use among astrologers.

The professors of this art recognised five planetary aspects, viz., opposition, conjunction, sextile, square, and trine, each possessing its peculiar kind of influence on events. The Moon, the planets, and the constellations in their conjunctions and configurations, were believed to reveal to those who could understand the significance of their aspects, the destiny of individuals and the occurrence of future events. The inauspicious influences of the heavenly bodies are described by Milton as contributing to the general disarrangement of the happy condition of things that existed before the Fall.

After having described the adverse physical changes which occurred in Nature as a consequence of the Fall, Milton makes use of his astronomical knowledge in explaining how they were brought about, and suggests two hypotheses : (1) a change of position of the Earth's axis ; (2) an alteration of the Sun's path from the equinoctial road :—

L

Some say he bid his Angels turn askance
The poles of Earth twice ten degrees and more
From the Sun's axle ; they with labour pushed
Oblique the centric globe : some say the Sun
Was bid turn reins from the equinoctial road
Like distant breadth—to Taurus with the seven
Atlantic Sisters, and the Spartan Twins,
Up to the Tropic Crab ; thence down amain
By Leo, and the Virgin, and the Scales,
As deep as Capricorn ; to bring in change
Of seasons to each clime.   Else had the spring
Perpetual smiled on Earth with vernant flowers.

x. 668-79

In support of the theory of a perpetual spring,
Milton assumes that the Earth's axis was directed
at right angles to her orbit, and that the plane of
the equator coincided with that of the ecliptic.
Consequently, the Sun's path remained always on
the equator, where his rays were vertical, and north
and south of this line each locality on the Earth
enjoyed one constant season, the character of which
depended upon its geographical position.   In what
are now the temperate regions of the globe there
was one continuous season, similar in climate and
length of day to what is experienced at the vernal
equinox, when the Sun is for a few days on the
equator.   There was then no winter, no summer,
nor autumn ; and, consequently, the growth of vegetation must have taken place under conditions of
climate entirely different to what exist on the
Earth at the present time.

The change of position of the Earth's axis,
'twice ten degrees and more from the Sun's axle,'

is described by Milton as having been accomplished by the might of angels, who ' with labour pushed oblique the centric globe.'

(2) According to the Ptolemaic belief, the Sun revolved round the Earth, but his course was altered from the equinoctial road to the path that he now pursues, which is the ecliptic. Instead of remaining on the equator, he travels an equal distance from this line upwards and downwards in each hemisphere.

The path of the Sun in the heavens is described by Milton with marked precision, and he mentions in regular order the names of the zodiacal constellations through which the orb travels. Passing through Taurus with the seven Atlantic Sisters (the Pleiades) and the Spartan Twins (Gemini), he enters the Tropic Crab (Cancer), in which constellation he attains his highest northern altitude; thence downwards he travels through Leo, Virgo, and the Scales (Libra), as deep as Capricornus, reaching his lowest point of declination at the winter solstice; and were it not for this alteration of the Sun's path, the poet informs us that perpetual spring would have reigned upon the Earth.

Milton was evidently well acquainted with the astronomical reasons (the revolution of the Earth in her orbit and the obliquity of the ecliptic) by which the occurrence and regular sequence of the seasons can be explained.

The path of the Sun in the heavens; his upward and downward course from the equator; the names of the constellations through which the orb travels, and the periods of the year at which he enters them, were also familiar to him.

The grateful change of the seasons, and the varied aspects of nature peculiar to each, which give a charm and freshness to the rolling year, must have been to Milton a source of pleasure and delight, and have stimulated his poetic fancy.

His observation of natural phenomena, and his keen perception of the pleasing changes which accompany them, are described in the following lines :—

> As, when from mountain-tops the dusky clouds
> Ascending, while the north wind sleeps, o'erspread
> Heaven's cheerful face, the louring element
> Scowls o'er the darkened landskip snow or shower,
> If chance the radiant Sun, with farewell sweet,
> Extend his evening beam, the fields revive,
> The birds their notes renew, and bleating herds
> Attest their joy, that hill and valley rings.—ii. 488–95.

The ancient poets Virgil and Ovid describe the Earth as having been created in the spring ; and associated with this season, which

> to the heart inspires
> Vernal delight and joy—iv. 154–55,

were the Graces and the Hours, which danced hand in hand as they led on the eternal Spring.

Milton alludes to the seasons on several occasions throughout his poem, and to the natural phenomena associated with them :—

> As bees
> In springtime when the Sun with Taurus rides,
> Pour forth their populous youth about the hive

In clusters; they among fresh dews and flowers
Fly to and fro, or on the smoothèd plank
The suburb of their straw-built citadel
New rubbed with balm, expatiate and confer
Their state affairs.—i. 768-75.

The Sun is in the constellation Taurus in April, when the warmth of his rays begins to impart new life and activity to the insect world after their long winter's sleep.

In his description of the repast partaken by the Angel Raphael with Adam and Eve in Paradise, Milton writes :—

Raised of grassy turf
Their table was, and mossy seats had round,
And on her ample square, from side to side,
All Autumn piled, though Spring and Autumn here
Danced hand in hand.—v. 391-95.

In describing Beelzebub when about to address the Stygian Council, he says :—

His look
Drew audience and attention still as night
Or summer's noontide air, while thus he spake.
ii. 307-309.

The failing vision from which Milton suffered in his declining years was succeeded by total blindness. This sad affliction he alludes to in the following lines :—

Thus with the year
Seasons return ; but not to me returns
Day, or the sweet approach of even or morn,
Or sight of vernal bloom, or summer's rose.—iii. 40-43.

We are able to perceive how much Milton was impressed with the beautiful seasons, and the varying aspects of the year which accompany them, and how his poetic imagination luxuriated in the changing variety of nature observable in earth and sky that from day to day afforded him exquisite delight; and, although his poem was written when blindness had overtaken him, yet those glad re-membrances remained as fresh in his memory as when in his youth he roamed among the flowery meadows, the vocal woodlands, and the winding lanes of Buckinghamshire.

The idea expressed by Milton that the primitive earth enjoyed a perpetual spring, though pleasing to the imagination, and well adapted for poetic description, is not sustained by any astronomical testimony. Indeed, the position of the Earth, with her axis at right angles to her orbit, is one which may be regarded as being ill adapted for the support and maintenance of life on her surface, just as her present position is the best that can be imagined for fulfilling this purpose.

Astronomy teaches us to rely with certainty upon the permanence and regular sequence of the seasons. The position of the Earth's axis as she speeds along in her orbit through the unresisting ether remains unchanged, and her rapid rotation has the effect of increasing its stability. Yet, the Earth performs none of her motions with rigid precision, and there is a very slow alteration of the position of her axis occurring, which, if unchecked, would eventually produce a coincidence of the equator and the ecliptic. Instead of a succession of the seasons, there would then be perpetual spring upon the Earth, and, although it would require a great epoch of time to bring about such a change, there would result a condition of things entirely different to what now exists on the globe. But, before the ecliptic can have approached sufficiently near the equator to produce any appreciable effect upon the climate of the Earth, its motion must cease, and after remaining stationary for a time, it will begin to recede to its former position. The seasons must therefore follow each other in regular sequence, and throughout all time, reminding us of the promise of the Creator, 'that while the Earth remaineth seed-time and harvest, and cold and heat, and summer and winter shall not cease.'

CHAPTER VI

THE STARRY HEAVENS

THE celestial vault, that, like a circling canopy of
sapphire hue, stretches overhead from horizon to
horizon, resplendent by night with myriad stars of
different magnitudes and varied brilliancy, forming
clusterings and configurations of fantastic shape
and beauty, arrests the attention of the most casual
observer.   But to one who has studied the heavens,
and followed the efforts of human genius in un-
ravelling the mysteries associated with those bright
orbs, the impression created on his mind as he
gazes upon them in the still hours of the night,
when the turmoil of life is hushed in repose, is one
of wonder and longing to know more of their being
and the hidden causes which brought them forth.
Here, we have poetry written in letters of gold on
the sable vestment of night; music in the gliding
motion of the spheres ; and harmony in the orbital
sweep of sun, planet, and satellite.

Milton was not only familiar with ' the face of
the sky,' as it is popularly called, but also knew the
structure of the celestial sphere, and the great
circles by which it is circumscribed.   Two of those
—the colures—he alludes to in the following lines,
when he describes the manner in which Satan, to

avoid detection, compassed the Earth, after his discovery by Gabriel in Paradise, and his flight thence :—

> The space of seven continued nights he rode
> With darkness—thrice the equinoctial line
> He circled, four times crossed the car of night
> From pole to pole, traversing each colure.—ix. 63–66.

Aristarchus of Samos believed the stars were golden studs, that illumined the crystal dome of heaven ; but modern research has transformed this conception of the ancient astronomer's into a universe of blazing suns rushing through regions of illimitable space. In Milton's time astronomers had arrived at no definite conclusion with regard to the nature of the stars. They were known to be self-luminous bodies, situated at a remote distance in space, but it had not been ascertained with any degree of certainty that they were suns, resembling in magnitude and brilliancy our Sun. Indeed, little was known of those orbs until within the past hundred years, when the exploration of the heavens by the aid of greatly increased telescopic power, was the means of creating a new branch of astronomical science, called sidereal astronomy.

We are indebted to Sir William Herschel, more than to any other astronomer, for our knowledge of the stellar universe. It was he who ascertained the vastness of its dimensions, and attempted to delineate its structural configuration. He also explored the star depths, which occupy the infinitude of space by which we are surrounded, and made many wonderful discoveries, which testify to his ability as an observer, and to his greatness as an astronomer.

William Herschel was born at Hanover, November 15, 1738. His father was a musician in the band of the Hanoverian Guard, and trained his son in his own profession. After four years of military service, young Herschel arrived in England when nineteen years of age, and maintained himself by giving lessons in music. We hear of him first at Leeds, where he followed his profession, and instructed the band of the Durham Militia. From Leeds he went to Halifax, and was appointed organist there ; on the expiration of twelve months he removed to Bath, and was elected to a similar post at the Octagon Chapel in that city. Here, fortune smiled upon him, and he became a busy and prosperous man. Besides attending to his numerous private engagements, he organised concerts, oratorios, and other public musical entertainments, which gained him much popularity among the cultivated classes which frequented this fashionable resort. Notwithstanding his numerous professional engagements, Herschel was able to devote a portion of his time to acquiring knowledge on other subjects. He became proficient in Italian and Greek, studied mathematics, and read books on astronomy. In 1773 he borrowed a small telescope, which he used for observational purposes, and was so captivated with the appearances presented by the celestial bodies, that he resolved to dedicate his life to acquiring 'a knowledge of the construction of the heavens.' This resolution he nobly adhered to, and became one of the most distinguished of astronomers. Like many other astronomers, Herschel possessed the requisite skill which enabled him to construct his own telescopes. Being desirous of possessing a more powerful instrument, and not having the means to purchase one, he commenced the manufacture of specula, the grinding and polishing of which had to be done by hand, entailing the necessity of tedious labour and the exercise of much patience. After repeated failures he at length completed a $5\frac{1}{2}$-foot Gregorian reflector, and with this instrument made his first survey of the heavens. Having perceived the desirability of possessing a more powerful telescope, he equipped himself with a reflector of twenty feet focal length, and it was with this instrument that he made those wonderful discoveries which established his reputation as a great astronomer.

On March 31, 1781, when examining the stars in the constellation Gemini, Herschel observed a star which presented an appearance slightly different to that of the other stars by which it was surrounded; it looked larger, had a perceptible disc, and its light became fainter when viewed with a higher magnifying power. After having carefully examined this object, Herschel arrived at the conclusion that he had discovered a comet. He communicated intelligence of his discovery to the Royal Society, and, a notification of it having been sent to

the Continental observatories, this celestial visitor
was subjected to a close scrutiny ; its progressive
motion among the stars was carefully observed, and
an orbit was assigned to it. After it had been
under observation for some time, doubts were
expressed as to its being a comet, these were increased on further examination, and eventually it
was discovered that this interesting object was a
new planet. This important discovery at once
raised Herschel to a position of eminence and distinction, and from a star-gazing musician he became a famous astronomer. A new planet named
Uranus was added to our system, which completes
a revolution round the Sun in a little over eighty-
four years, and at a distance of near 1,000 millions
of miles beyond the orbit of Saturn. Herschel's
name became a household word. George III. invited him to Court in order that he might obtain
from his own lips an account of his discovery of the
new planet ; and so favourable was the impression
made by Herschel upon the King, that he proposed
to create him Royal Astronomer at Windsor, and
bestow upon him a salary of 200l. a year. Herschel decided to accept the proffered appointment,
and, with his sister Caroline, removed from Bath to
Datchet, near Windsor, in 1782, and from there to
Slough in 1786. In 1788 he married the wealthy
widow of a London merchant, by whom he had one
son, who worthily sustained his father's high reputation as an astronomer. Herschel was created a
Knight in 1816, and in 1821 was elected first

President of the Royal Astronomical Society. He died at Slough on August 25, 1822, when in the eighty-fourth year of his age, and was buried in Upton Churchyard.

It is inscribed on his tomb, that 'he burst the barriers of heaven;' the lofty praise conveyed by this expression is not greater than what Herschel merited when we consider with what unwearied assiduity and patience he laboured to accomplish the results described in the words which have been quoted. By a method called 'star-gauging' he accomplished an entire survey of the heavens and examined minutely all the stars in their groups and aggregations as they passed before his eye in the field of the telescope. He sounded the depths of the Milky Way, and explored the wondrous regions of that shining zone, peopled with myriads of suns so closely aggregated in some of its tracts as to suggest the appearance of a mosaic of stars. He resolved numerous nebulæ into clusters of stars, and penetrated with his great telescope depth after depth of space crowded with 'island universes of stars,' beyond which he was able to discern luminous haze and filmy streaks of light, the evidence of the existence of other universes plunged in depths still more profound, where space verges on infinity. In his exploration of the starry heavens Herschel's labours were truly amazing. On four different occasions he completed a survey of the firmament, and counted the stars in several thousand gauge-fields; he discovered 2,400 nebulæ,

800 double stars, and attempted to ascertain the approximate distances of the stars by a comparison of their relative brightness.

It had long been surmised, though no actual proof was forthcoming, that the law of gravitation by which the order and stability of our system are maintained exercises its potent influence over other material bodies existing in space, and that other systems, though differing in many respects from that of ours, and presenting a more complex arrangement in their structure, perform their motions subject to the guidance of this universal law. The uncertainty with regard to the controlling influence of gravity was removed by Herschel when he made his important discovery of binary star systems. The components of a binary star are usually in such close proximity that, to the naked eye, they appear as one star, and sometimes, even with telescopic aid, it is impossible to distinguish them individually; but when observed with sufficient magnifying power they can be easily perceived as two lucid points. Double stars were for a long time believed to be a purely optical phenomenon—an effect created by two stars projected on the sphere so as to appear nearly in the same line of vision, and, although apparently almost in contact, situated at great distances apart. At one time Herschel entertained a similar opinion with regard to those stars. In 1779 he undertook an extensive exploration of the heavens with the object of discovering double stars. As a result of his labours he presented to the

Royal Society in 1782 a list of 269 newly discovered double stars, and in three years after he sup-
plemented this list with another which contained
434 more new stars. He carefully measured the
distances by which the component stars were
separated, and determined their position angles, in
order that he might be able to detect the existence
of any sensible parallax. On repeating his observations twenty years after, he discovered that
the relative positions of many of the stars had
changed, and in 1802 he made the important announcement of his discovery that the components
of many double stars form independent systems,
held together in a mutual bond of union and revolving round one common centre of gravity.

The importance of this discovery, which we owe
to Herschel's sagacity and accuracy of observation,
cannot be over-estimated ; what was previously conjecture and surmise, now became precise knowledge
established upon a sure and accurate basis. It
was ascertained that the law of gravity exerts its
power in regulating and controlling the motions of
all celestial bodies within the range of telescopic
vision, and that the order and harmony which
pervade our system are equally present among
other systems of suns and worlds distributed
throughout the regions of space. The spectacle
of two or more suns revolving round each other,
forming systems of greater magnitude and importance than that of ours, conveyed to the minds
of astronomers a knowledge of the mechanism of

the heavens which had hitherto been unknown to them.

During the many years which Herschel devoted to the exploration of the starry heavens, and when engaged night after night in examining and enumerating the various groups and clusters of stars which passed before his eye in the field of his powerful telescope, he did not fail to remember the sublime object of his life, and to which he made all his other investigations subordinate, viz., the delineation of the structural configuration of the heavens, and the inclusion of all aggregations, groups, clusters, and galaxies of stars which are apparently scattered promiscuously throughout the regions of space into one grand harmonious design of celestial architecture.

Having this object in view, he explored the wondrous zone of the Milky Way, gauged its depths, measured its dimensions, and, in attempting to unravel the intricacies of its structure, penetrated its recesses far beyond the limit attained by any other observer. Acting on the assumption that the stars are uniformly distributed throughout space, Herschel, by his method of star-gauging, concluded that the sidereal system consists of an irregular stratum of evenly distributed suns, resembling in form a cloven flat disc, and that the apparent richness of some regions as compared with that of others could be accounted for by the position from which it was viewed by an observer. The stars would appear least numerous where the

visual line was shortest, and, as it became lengthened,
they would increase in number until, by crowding
behind each other as a greater depth of stratum
was penetrated, they would, when very remote,
present the appearance of a luminous cloud or zone
of light. After further observation Herschel was
compelled to relinquish his theory of equal star dis-
tribution, and found, as he approached the Galaxy,
that the stars became much more numerous, and
that in the Milky Way itself there was evidence of the
gravitation of stars towards certain regions forming
aggregations and clusters which would ultimately
lead to its breaking up into numerous separate
sidereal systems. As he extended his survey of the
heavens and examined with greater minuteness the
stellar regions in the Galactic tract, he discovered
that by his method of star-gauging he was unable
to define the complexity of structure and variety of
arrangement which came under his observation ; he
also perceived that the star-depths are unfathomable,
and discerned that beyond the reach of his telescope there existed systems and galaxies of stars
situated at an appalling distance in the abysmal
depths of space. Though the magnitude of that
portion of the sidereal heavens which came under his
observation was inconceivable as regards its dimensions, Herschel was able to perceive that it formed
but a part—and most probably a small part—of the
stellar universe, and that without a more extended
knowledge of this universe, which at present is un-
attainable, it would be impossible to determine its

M

structural configuration or discover the relationships that exist among the sidereal systems and Galactic concourses of stars distributed throughout space. Herschel ultimately abandoned his star-gauging method of observation and confined his attention to exploring the star depths and investigating the laws and theories associated with the bodies occupying those distant regions.

Since all the planets if viewed from the Sun would be seen to move harmoniously and in regular order round that body, so there may be somewhere in the universe a central point, or, as some persons imagine, a great central sun, round which all the systems of stars perform their majestic revolutions with the same beautiful regularity; having their motions controlled by the same law of gravitation, and possessing the same dynamical stability which characterises the mechanism of the solar system.

The extent of the distance which intervenes between our system and the fixed stars constituted a problem which exercised the minds of astronomers from an early period until the middle of the present century.

Tycho Brahé, who repudiated the Copernican theory, asserted as one of his reasons against it that the distances by which the heavenly bodies are separated from each other were greater than even the upholders of this theory believed them to be. Although the distance of the Sun from the Earth was unknown, Tycho was aware that the diameter of the Earth's orbit must be measured by millions of miles, and yet there was no perceptible motion or change of position of the stars when viewed from any point of the vast circumference which she traverses. Consequently, the Earth, if viewed from the neighbourhood of a star, would also appear motionless, and the dimensions of her orbit would be reduced to that of a point. This seemed incredible to Tycho, and he therefore concluded that the Copernican theory was incorrect.

The conclusion that the stars are orbs resembling our Sun in magnitude and brilliancy was one which, Tycho urged, should not be hastily adopted; and yet, if it were conceded that the Earth is a body which revolves round the Sun, it would be necessary to admit that the stars are suns also. If the Earth's orbit, as seen from a star, were reduced to a point, then the Sun, which occupies its centre, would be reduced to a point of light also, and, when observed from a star of equal brilliancy and magnitude, would have the same resemblance that the star has when viewed from the Earth, which may be regarded as being in proximity to the Sun. Tycho Brahé would not admit the accuracy of these conclusions, which were too bewildering and overwhelming for his mental conception.

But the investigations of later astronomers disclosed the fact that the heavenly bodies are situated at distances more remote from each other than had been previously imagined, and that the reasons which led Tycho to reject the Copernican theory were based upon erroneous conclusions, and could,

with greater aptitude, be employed in its support.
It was ascertained that the distance of the Sun from
the Earth, which at different periods was surmised
to be ten, twenty, and forty millions of miles, was
much greater than had been previously estimated.
Later calculations determined it to be not less than
eighty millions of miles, and, according to the most
recent observations, the distance of the Sun from
the Earth is believed to be about ninety-three
millions of miles.

Having once ascertained the distance between
the Earth and the Sun, astronomers were enabled
to determine with greater facility the distances of
other heavenly bodies.

It was now known that the diameter of the
Earth's orbit exceeded 183 millions of miles, and
yet, with a base line of such enormous length, and
with instruments of the most perfect construction,
astronomers were only able to perceive the minutest
appreciable alteration in the positions of a few stars
when observed from opposite points of the terrestrial orbit.

It had long been the ambitious desire of astronomers .to accomplish, if possible, a measurement
of the abyss which separates our system from the
nearest of the fixed stars.   No imaginary measuring
line had ever been stretched across this region of
space, nor had its unfathomed depths ever been
sounded by any effort of the human mind.   The
stars were known to be inconceivably remote, but
how far away no person could tell, nor did there

exist any guide by which an approximation of their distances could be arrived at.

In attempting to calculate the distances of the stars, astronomers have had recourse to a method called 'Parallax,' by which is meant the apparent change of position of a heavenly body when viewed from two different points of observation.

The annual parallax of a heavenly body is the angle subtended at that body by the radius of the Earth's orbit.

The stars have no diurnal parallax, because, owing to their great distance, the Earth's radius does not subtend any measurable angle, but the radius of the Earth's orbit, which is immensely larger, does, in the case of a few stars, subtend a very minute angle.

'This enormous base line of 183 millions of miles is barely sufficient, in conjunction with the use of the most delicate and powerful astronomical instruments, to exhibit the minutest measureable displacement of two or three of the nearest stars.' —Proctor.

The efforts of early astronomers to detect any perceptible alteration in the positions of the stars when observed from any point of the circumference of the Earth's orbit were unsuccessful. Copernicus ascribed the absence of any parallax to the immense distances of the stars as compared with the dimensions of the terrestrial orbit. Tycho Brahé, though possessing better appliances, and instruments of more perfect construction, was unable to perceive any annual displacement of the stars, and brought this forward as evidence against the Copernican theory.

Galileo suggested a method of obtaining the parallax of the fixed stars, by observing two stars of unequal magnitude apparently near to each other, though really far apart. Those, when observed from different points of the Earth's orbit, would appear to change their positions relatively to each other. The smaller and more distant star would remain unaltered, whilst the larger and nearer star would have changed its position with respect to the other. By continuing to observe the larger star during the time that the Earth accomplished a revolution of her orbit, Galileo believed that its parallax might be successfully determined. Though he did not himself put this method into practice, it has been tried by others with successful results.

In 1669, Hooke made the first attempt to ascertain the parallax of a fixed star, and selected for this purpose γ Draconis, a bright star in the Head of the Dragon. This constellation passed near the zenith of London at the time that he made his observations, and was favourably situated, so as to avoid the effects of refraction. Hooke made four observations in the months of July, August, and October, and believed that he determined the parallax of the star; but it was afterwards discovered that he was in error, and that the apparent displacement of the star was mainly due to the aberration of light—a phenomenon which was not discovered at that time.

A few years later, Picard, a French astronomer, attempted to find the parallax of a Lyrae, but was unsuccessful. In 1692–93, Roemer, a Danish astronomer, observed irregularities in the declinations of the stars which could neither be ascribed to parallax or refraction, and which he imagined resulted from a changing position of the Earth's axis.

One of the principal causes which baffled astronomers in their endeavours to determine the parallax of the fixed stars was a phenomenon called the ' Aberration of Light,' which was discovered and explained by Bradley in 1727. The peculiar effect of aberration was perceived by him when endeavouring to obtain the parallax of γ Draconis.

Owing to the progressive transmission of light, conjointly with the motion of the Earth in her orbit, there results an apparent slight displacement of a star from its true position. The extent of the displacement depends upon the ratio of the velocity of light as compared with the speed of the Earth in her orbit, which is as 10,000 to 1. As a consequence of this, each star describes a small ellipse in the course of a year, the central point of which would indicate the place occupied by the star if the Earth were at rest. The shifting position of the star is very slight, and at the end of a year it returns to its former place.

Prior to the discovery of aberration, astronomers ascribed the apparent displacement of the stars arising from this cause as being due to parallax—a conclusion which led to erroneous results ; but after Bradley's discovery this source of error was avoided, and it was found that the parallax of the stars had to be considerably reduced.

Bessel was the first astronomer who merited the high distinction of having determined the first reliable stellar parallax, and by this achievement he was enabled to fathom the profound abyss which separates our solar system from the stars.

Frederick William Bessel was born in 1764 at Minden, in Westphalia. It was his intention to pursue a mercantile career, and he commenced life by becoming apprenticed to a firm of merchants at Bremen. Soon afterwards he accompanied a trading expedition to China and the East Indies, and while on this voyage picked up a good deal of information with regard to many matters which came under his observation. He acquired a knowledge of Spanish and English, and made himself acquainted with the art of navigation. On his return home, Bessel endeavoured to determine the longitude of Bremen. The only appliances which he made use of were a sextant constructed by himself, and a common clock ; and yet, with those rude instruments, he successfully accomplished his object. During the next two years he devoted all his spare time to the study of mathematics and astronomy, and, having obtained possession of Harriot's observations of the celebrated comet of 1607—known as Halley's comet—Bessel, after much diligent application and careful calculation, was enabled to deduce from them an orbit, which he assigned to that remarkable body. This meritorious achievement was the means of procuring for him a widely known reputation.

A vacancy for an assistant having occurred at Schröter's Observatory at Lilienthal, the post was offered to Bessel and accepted by him. Here he remained for four years, and was afterwards appointed Director of the new Prussian Observatory at Königsberg, where he pursued his astronomical labours for a period of upwards of thirty years. Bessel directed his energies chiefly to the study of stellar astronomy, and made many observations in determining the number, the exact positions, and proper motions of the stars. He was remarkable for the precision with which he carried out his observations, and for the accuracy which characterised all his calculations.

In 1837 Bessel, by the exercise of his consummate skill, endeavoured to solve a problem which for many years baffled the efforts of the ablest astronomers, viz., the determination of the parallax of the fixed stars. This had been so frequently attempted, and without success, that the results of any new observations were received with incredulity before their value could be ascertained.

Bessel was ably assisted by Joseph Frauenhofer, an eminent optician of Munich, who constructed a magnificent heliometer for the Observatory at Königsberg, and in its design introduced a principle

which admirably adapted it for micrometrical measurement.

The star selected by Bessel is a binary known
as 61 Cygni, the components being of magnitudes
5·5 and 6 respectively. It has a large proper
motion, which led him to conclude that its parallax must be considerable.

This star will always be an object of interest to
astronomers, as it was the first of the stellar multitude that revealed to Bessel the secret of its distance.

Bessel commenced his observations in October
1837, and continued them until March 1840.
During this time he made 402 measurements, and,
before arriving at a conclusive result, carefully considered every imaginable cause of error, and rigor-
ously calculated any inaccuracies that might arise
therefrom. Finally, he determined the parallax of
the star to be 0″·3483—a result equivalent to a distance about 600,000 times that of the Earth from
the Sun. In 1842–43 M. Peters, of the Pulkova
Observatory, arrived at an almost similar result,
having obtained a parallax of 0″·349 ; but by more
recent observations the parallax of the star has
been increased to about half a second.

About the same time that Bessel was occupied
with his observation of 61 Cygni, Professor Henderson, of Edinburgh, when in charge of the Observatory at the Cape of Good Hope, directed his
attention to α Centauri, one of the brightest stars
in the Southern Hemisphere. During 1832–33 he
made a series of observations of the star, with the

object of ascertaining its mean declination; and, having been informed afterwards of its large proper motion, he resolved to make an endeavour to determine its parallax. This he accomplished after his return to Scotland, having been appointed Astronomer Royal in that country. By an examination of the observations made by him at the Cape, he determined the parallax of α Centauri to be $1''\cdot16$, but later astronomers have reduced it to $0''\cdot75$.

Professor Henderson's detection of the parallax of α Centauri was communicated to the Astronomical Society two months after Bessel announced his determination of the parallax of 61 Cygni.

The parallax of 61 Cygni assigns to the star a distance of forty billions of miles from the Earth, and that of α Centauri—regarded as the nearest star to our system—a distance of twenty-five billions of miles.

It is utterly beyond the capacity of the human mind to form any adequate conception of those vast distances, even when measured by the velocity with which the ether of space is thrilled into light. Light, which travels twelve millions of miles in a minute, requires $4\frac{1}{3}$ years to cross the abyss which intervenes between α Centauri and the Earth, and from 61 Cygni the period required for light to reach our globe is rather less than double that time.

The parallax of more than a dozen other stars has been determined, and the light passage of a few of the best known is estimated as follows:—
Sirius, eight years; Procyon, twelve; Altair, six-

teen; Aldebaran, twenty-eight; Capella, thirty; Regulus, thirty-five; Polaris, sixty-three; and Vega, ninety-six years.

It does not always follow that the brightest stars are those situated nearest to our system, though in a general way this may be regarded as correct. The diminishing magnitudes of the stars can be accounted for mainly by their increased distances, rather than by any difference in their intrinsic brilliancy. We should not err by inferring that the most minute stars are also the most remote; the telescope revealing thousands that are invisible to the naked eye. There are, however, exceptions to this general rule, and there are many stars of small magnitude less remote than those whose names have been enumerated, and whose light passage testifies to their profound distances and surpassing magnitude when compared with that of our Sun.

Sirius, 'the leader of the heavenly host,' is distant fifty billions of miles. The orb shines with a brilliancy far surpassing that of the Sun, and greatly exceeds him in mass and dimensions. Arcturus, the bright star in Boötes, whose golden yellow light renders it such a conspicuous object, is so far distant that its measurement gives no reliable parallax; and if we may infer from what little we know of the stars, Arcturus is believed to be the most magnificent and massive orb entering into the structure of that portion of the sidereal system which comes within our cognisance. Judging by its relative size and brightness, this star is ten thousand times more luminous, and may exceed the Sun one million times in volume.

Deneb, in the constellation of the Swan, though a first-magnitude star, possesses no perceptible proper motion or parallax—a circumstance indicative of amazing distance, and magnitude equalling, or surpassing, Arcturus and Sirius.

Canopus, in the constellation Argo, in the Southern Hemisphere, the brightest star in the heavens with the exception of Sirius, possesses no sensible parallax; consequently, its distance is unknown, though it has been estimated that its light passage cannot be less than sixty-five years.

By establishing a mean value for the parallax of stars of different magnitudes, it was believed that an approximation of their distances could be obtained by calculating the time occupied in their light passage. The light period for stars of the first magnitude has been estimated at thirty-six and a half years ; this applies to the brightest stars, which are also regarded as the nearest. At the distance indicated by this period, the Sun would shrink to the dimensions of a seventh-magnitude star and become invisible to the naked eye ; this of itself affords sufficient proof that the great luminary of our system cannot be regarded as one of the leading orbs of the firmament. Stars of the second magnitude have a mean distance of fifty-eight light years, those of the third magnitude ninety-two years, and so on. M. Peters estimated that light from stars of the sixth magnitude, which are just visible to the naked eye, requires a period of 138 years to accomplish its journey hither; whilst light emitted from the smallest stars visible in large telescopes does not reach the Earth until after the lapse of thousands of years from the time of leaving its source.

The profound distances of the nearest stars by which we are surrounded lead us to consider the isolated position of the solar system in space. A pinnacle of rock, or forsaken raft floating in mid-ocean, is not more distant from the shore than is the Sun from his nearest neighbours. The inconceivable dimensions of the abyss by which the orb and his attendants are surrounded in utter loneliness may be partially comprehended when it is known that light, which travels from the Sun to the Earth—a distance of ninety-three millions of miles —in eight minutes, requires a period of four and a third years to reach us from the nearest fixed star. A sphere having the Sun at its centre and this nearest star at its circumference would have a diameter of upwards of fifty billions of miles; the volume of the orb when compared with the dimensions of this circular vacuity of space is as a small shot to a globe 900 miles in diameter. It has been estimated by Father Secchi that, if a comet when at aphelion were to arrive at a point midway between the Sun and the nearest fixed star, it would require one hundred million years in the accomplishment of its journey thither. And yet the Sun is one of a group of stars which occupy a region of the heavens adjacent to the Milky Way and surrounded by that zone ; nor is his isolation greater than that of those stars which are his companions, and who, notwithstanding their profound distance, influence his movements by their gravitational attraction, and in combination with the other stars of the firmament control his destiny.

Ancient astronomers, for the purpose of description, have mapped out the heavens into numerous irregular divisions called ' constellations.' They are of various forms and sizes, according to the configuration of the stars which occupy them, and have been named after different animals, mythological heroes, and other objects which they appear to resemble. In a few instances there does exist a similitude to the object after which a constellation is called ; this is evident in the case of Corona Borealis (the Northern Crown), in which there can be seen a conspicuous arrangement of stars resembling a coronet, and in the constellations of the Dolphin and Scorpion, where the stars are so distributed that the forms of those creatures can be readily recognised. There is some slight resemblance to a bear in Ursa Major, and to a lion in Leo, and no great effort of the mind is required to imagine a chair in Cassiopeia, and a giant in Orion ; but in the majority of instances it is difficult to perceive any likeness of the object after which a constellation is named, and in many cases there is no resemblance whatever.

The constellations are sixty-seven in number : excluding those of the Zodiac, which have been already mentioned, the constellations of the Northern Hemisphere number twenty-nine. The most important of these are Ursa Major and Minor, Andromeda, Cassiopeia, Cepheus, Cygnus, Lyra, Aquila, Auriga, Draco, Boötes, Hercules, Pegasus, and Corona Borealis.

To an observer of the nocturnal sky the stars appear to be very unequally distributed over the celestial sphere. In some regions they are few in number and of small magnitude, whilst in other parts of the heavens, and especially in the vicinity of the Milky Way, they are present in great numbers and form groups and aggregations of striking appearance and conspicuous brilliancy. On taking a casual glance at the midnight sky on a clear moonless night, one is struck with the apparent countless multitude of the stars ; yet this impression of their vast number is deceptive, for not more than two thousand stars are usually visible at one time.

Much, however, depends upon the keenness of vision of the observer, and the transparency of the atmosphere. Argelander counted at Bonn more than 3,000 stars, and Hozeau, near the equator, where all the stars of the sphere successively appear in view, enumerated 6,000 stars. This number may be regarded as including all the stars in the heavens that are visible to the naked eye. With the aid of an opera glass thousands of stars can be seen

that are imperceptible to ordinary vision. Argelander, with a small telescope of $2\frac{1}{2}$ inches aperture,
was able to count 234,000 stars in the Northern
Hemisphere. Large telescopes reveal multitudes of
stars utterly beyond the power of enumeration, nor
do they appear to diminish in number as depth
after depth of space is penetrated by powerful
instruments. The star-population of the heavens
has been reckoned at 100,000,000, but this estimate
is merely an assumption ; recent discoveries made
by means of stellar photography indicate that the
stars exist in myriads. It is reasonable to believe
that there is a limit to the sidereal universe, but it
is impossible to assign its bounds or comprehend
the apparently infinite extent of its dimensions.

Scintillation or twinkling of the stars is a property which distinguishes them from the planets.
It is due to a disturbed condition of the atmosphere
and is most apparent when a star is near the horizon ; at the zenith it almost entirely vanishes.
Humboldt states that in the clear air of Cumana,
in South America, the stars do not twinkle after
they reach an elevation of 15° above the horizon.
The presence of moisture in the atmosphere inten-
sifies scintillation, and this is usually regarded as a
prognostication of rain. White stars twinkle more
than red ones. The occurrence of scintillation can
be accounted for by the fact that the stars are
visible as single points of light which twinkle as a
whole, but in the case of the Sun, Moon, and planets,
they form discs from which many points of light

are emitted ; they, therefore, do not scintillate as a whole, for the absence of rays of light from one portion of their surface is compensated by those from other parts of their discs, giving a mean average which creates a steadiness of vision.

The stars are divided into separate classes called 'magnitudes,' by which their relative apparent size and degree of brightness are distinguished. The magnitude of a star does not indicate its mass or dimensions, but its light-giving power, which depends partly upon its size and distance, though mainly upon the intensity of its luminosity. The most conspicuous are termed stars of the first magnitude; there are ten of those in the Northern Hemisphere, and an equal number south of the equator, but they are not all of the same brilliancy. Sirius outshines every other star of the firmament, and Arcturus has no rival in the northern heavens. The names of the first-magnitude stars north of the equator are : Arcturus, Capella, Vega, Betelgeux, Procyon, Aldebaran, Altair, Pollux, Regulus, and Deneb. The next class in order of brightness are called second-magnitude stars; they are fifty or sixty in number, the most important of which is the Pole Star. The stars diminish in luminosity by successive gradations, and when they sink to the sixth magnitude reach the utmost limit at which they appear visible to the naked eye. In great telescopes this classification is carried so low as to include stars of the eighteenth and twentieth magnitudes.

Entering into the structure of the stellar universe we have Single Stars, Double Stars, Triple, Quadruple, and Multiple Stars, Temporary, Periodical, and Variable Stars, Star-groups, Star-clusters, Galaxies, and Nebulæ.

SINGLE OR INSULATED STARS include all those orbs sufficiently isolated in space so as not to be perceptibly influenced by the attraction of other similar bodies. They are believed to constitute the centres of planetary systems, and fulfil the purpose for which they were created by dispensing light and heat to the worlds which circle around them.

The Sun is an example of this class of star, and constitutes the centre of the system to which the Earth belongs. Reasoning from analogy, it would be natural to conclude that there are other suns, numberless beyond conception, the centres of systems of revolving worlds, and although we are utterly unable to catch a glimpse of their planetary attendants, even with the aid of the most powerful telescopes, yet they have in a few instances been *felt*, and have afforded unmistakable indications of their existence.

Since the Sun must be regarded as one of the stellar multitude that people the regions of space, and whose surpassing splendour when contrasted with that of other luminaries can be accounted for by his proximity to us, it would be of interest to ascertain his relative importance when compared with other celestial orbs which may be his peers or his superiors in magnitude and brilliancy.

The Sun is one of a widely scattered group of stars situated in the plane of the Milky Way and surrounded by that zone, and, as a star among the stars, would be included in the constellation of the Centaur.

Although regarded as one of the leading orbs of the firmament, and of supreme importance to us, astronomers are undecided whether to classify the Sun with stars of greater magnitude and brightness, or assign him a position among minor orbs of smaller size. Much uncertainty exists with regard to star magnitudes. This arises from inability on the part of astronomers to ascertain the distances of the vast majority of stars visible to the naked eye, and also on account of inequality in their intrinsic brilliancy. Among the stars there exists an indefinite range of stellar magnitudes. There are many stars known whose dimensions have been ascertained to greatly exceed those of the Sun, and there are others of much smaller size. No approximation of the magnitude of telescopic stars can be arrived at; many of them may rival Sirius, Canopus, and Arcturus, in size and splendour, their apparent minuteness being a consequence of their extreme remoteness. If the Sun were removed a distance in space equal to that of many of the brightest stars, he would in appearance be reduced to a minute point of light or become altogether invisible; and there are other stars, situated at distances still more remote, of which sufficient is known to justify us in arriving at the conclusion that the Sun must be ranked among the minor orbs of the firmament, and that many of the stars surpass him in brilliancy and magnitude.

DOUBLE STARS.—To the unaided eye, these appear as single points of light; but, when observed with a telescope of sufficient magnifying power, their dual nature can be detected.

The first double star discovered was Mizar, the middle star of the three in Ursa Major which form the tail of the bear. The components are of the fourth and fifth magnitudes, of a brilliant white colour, and distant fourteen seconds of arc.

In 1678, Cassini perceived stars which appeared as single points of light when viewed with the naked eye, but when observed with the telescope presented the appearance of being double.

The astronomer Bode, in 1781, published a list of eighty double stars, and, in a few years after, Sir William Herschel discovered several hundreds more of those objects. They are now known to exist in thousands, Mr. Burnham, of the Lick Observatory, having, by his keen perception of vision, contributed more than any other observer to swell their number.

All double stars are not binaries; many of them are known as ' optical doubles '—an impression created by two stars when almost in the same line of vision, and, though apparently near, are situated at a great distance apart and devoid of any physical relationship.

Binary stars consist of two suns which revolve

round their common centre of gravity, and form real dual systems.

The close proximity of the components of double stars impressed the minds of some astronomers with the belief that a physical bond of union existed between them.   In the interval between 1718 and 1759, Bradley detected a change of 30° in the position angle of the two stars forming Castor, and was very nearly discovering their physical connection.

In 1767, the Rev. John Michell wrote : ' It is highly probable in particular, and next to a certainty in general, that such double stars as appear to consist of two or more stars placed very near together do really consist of stars placed near together and under the influence of some general law.'   Afterwards he says : ' It is not improbable that a few years may inform us that some of the great number of double and triple stars which have been observed by Mr. Herschel are systems of bodies revolving about each other.'   Christian Mayer, a German astronomer, formed a list of stellar pairs, and announced, in 1776, the supposed discovery of ' satellites ' to many of the principal stars.   His observations were, however, not exact enough to lead to any useful results, and the existence of his ' planet stars ' was at that time derided, and believed to find a place only in his imagination.

The conclusions arrived at by some astronomers with regard to double stars were afterwards confirmed by Herschel, when, by his observation of a change in the relative positions of many of their components, he was able to announce that they form independent systems in mutual revolution, and are controlled by the law of gravitation.

The number of binary stars in active revolution is known to exceed 500; but, besides these, there are doubtless numerous other compound stars which, on account of their extreme remoteness and the close proximity of their components, are irresolvable into pairs by any optical appliances which we possess.

The revolution of two suns in one sphere presents to our observation a scheme of creative design entirely different to the single-star system with which we are familiar—one of a higher and more complex order in the ascending scale of celestial architecture. For, if we assume that around each revolving sun there circles a retinue of planetary worlds, it is obvious that a much more complicated arrangement must exist among the orbs which enter into the formation of such a system than is found among those which gravitate round our Sun.

The common centre of gravity of a binary system is situated on a line between both stars, and distant from each in inverse proportion to their respective masses. When the stars are of equal mass their orbits are of equal dimensions, but when the mass of one star exceeds that of the other, the orbit of the larger star is proportionately diminished as compared with the circumference traversed by the smaller star. When their orbits are circular—a rare

occurrence—both stars pursue each other in the
same path, and invariably occupy it at diametrically
opposite points ; nor is it possible for one star to
approach the other by the minutest interval of space
in any duration of time, so long as the synchronous
harmony of their revolution remains undisturbed.

When a pair of suns move in an ellipse, their
orbits intersect and are of equal dimensions when

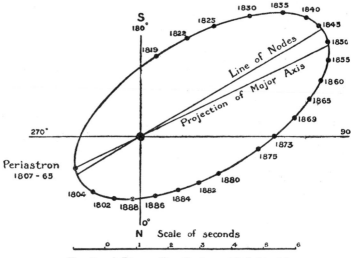

FIG. 3.—A Binary Star System—70 Ophiuchi
(*Drawn by Mr. J. E. Gore.*)

the stars are of equal mass, their common centre of
gravity being then at a point equidistant from each.
Consequently, neither star can approach or recede
from this point without the other affecting a similar
motion, they must be at periastron and apastron
together, and any acceleration or retardation of
speed must occur simultaneously with each. Stars
of unequal magnitude always maintain a propor-

tionate distance from their common focus, and both simultaneously occupy corresponding parts of their orbits.

The nature of the motions of those distant suns, and the form of the orbits which they traverse, have been investigated by several eminent astronomers, and although the subject is one of much difficulty, on account of their extreme remoteness and the minute angles which have to be dealt with, necessitating the carrying out of very refined observations, yet a considerable amount of information has been obtained with regard to the paths which they pursue in the accomplishment of their revolutions round each other.

The orbits of about sixty stellar pairs have been computed, but only with partial success. Some stars have shown themselves to be totally regardless of theory and computation, and have shot ahead far beyond the limits ascribed to them, whilst others, by the slowness of their motions, have upset the calculations of astronomers as much in the opposite direction. So that out of this number the orbits of not more than half a dozen are satisfactorily known.

The dimensions of stellar orbits are of very varied extent. Some pairs are apparently so close that the best optical means which we possess are incapable of dividing them, whilst others revolve in wide and spacious orbits.

The most marked peculiarity of the orbits of binary stars is their high eccentricity ; they are usually much more eccentric than are those of the planets, and in some instances approach in form that of a comet.

The finest binary star in the northern heavens is Castor, the brighter of the two leading stars in the constellation Gemini. The components are of the second and third magnitudes, and over five seconds apart. They are of a brilliant white colour, and form a beautiful object in the telescope.

In 1719 Bradley determined the relative positions of those stars, and on comparing the results obtained by him with recent measurements it was found that they had altered to the extent of 125°. Travelling at the same rate of speed, they will require a period of about 420 years to complete an entire circuit of their orbits. This pace, however, has not been maintained, for, their periastron having occurred in 1750, they travelled more rapidly in the last century than they are doing at present, and, as their orbits are so eccentric that when at apastron the stars are twice as remote from each other as at periastron, they will for the next three and a half centuries continue to slacken their pace, until they shall have reached the most remote points of their orbits, when they will again begin to approach with an increasing velocity; so that the time in which an entire revolution can be accomplished will not be much less than 1,000 years.[1]

As the distance of Castor is unknown, it is impossible to compute the combined mass of its com-

---

[1] Miss Clerke's *System of the Stars.*

ponents. They are very remote, their light period being estimated at forty-four years. Castor is doubtless a more massive orb than our Sun, and possesses a higher degree of luminosity.

a Centauri, in the Southern Hemisphere, is the brightest binary, and also the nearest known star in the heavens ; its estimated distance being twenty-five billions of miles. Both components equal stars of the first magnitude, and are of a brilliant white colour. Since they were first observed, in 1709, they have completed two revolutions, and are now accomplishing a third. The eccentricity of their orbit approaches in form that of Faye's comet, which travels round the Sun ; consequently the stars, when at apastron, are twice their periastron distance. Their period of revolution is about eighty-eight years. The mean radius of their orbit corresponds to a span of 1,000 millions of miles, so that those orbs are sometimes as close to each other as Jupiter is to the Sun, and never so far distant as Uranus.[1] Their combined mass is twice that of the Sun, and the luminosity of each star is slightly greater.

The double star 61 Cygni—one of the nearest to our system—is believed to be a binary the components of which move in an orbit of more spacious dimensions than that of any other known revolving pair. Though they have been under continuous observation since 1753, it is only within the last few years that any orbital motion has been perceived.

---

[1] Miss Clerke's *System of the Stars.*

Some observers are disinclined to admit the accuracy of this statement; whilst others believe that the stars have executed a hyperbolic sweep round their common centre of gravity and are now separating.

The radius of the orbit in which those bodies travel is sixty-five times the distance of the Earth from the Sun; which means that they travel in an orbit twice the width of that of the planet Neptune. It has been estimated that they complete a revolution in about eight centuries. The united mass of the system is about one-half that of the Sun, and in point of luminosity they are much inferior to that orb.

The star 70 Ophiuchi (fig. 3) may be regarded as typical of a binary system. The components are five seconds apart, and of the fourth and sixth magnitudes. Their light period is stated to be twenty years, and the combined mass of the system is nearly three times that of the Sun. The pair travel in an orbit from fourteen to forty-two times the radius of the Earth's orbit; so that when at apastron they are three times as distant from each other as when at periastron. They complete a revolution in eighty-eight years.

The accompanying diagram (fig. 4) is a delineation of the beautiful orbits of the components of γ Virginis. These may be described as elongated ellipses. Both stars being of equal mass, their orbits are of equal dimensions, and their common centre of gravity at a point equidistant from each.

Any approach to, or recession from this point, must occur simultaneously with each; they must always occupy corresponding parts of their orbits, and be in apastron and at periastron in the same period of time. The ellipse described by this pair is the most eccentric of known binary orbits, and approaches

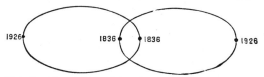

FIG. 4.—The Orbits of the Components of γ Virginis.

n form the path pursued by Encke's comet round the Sun. These orbs complete a revolution in 180 years, and when in apastron are seventeen times more remote from each other than when at periastron.

From his observation of the motion of Sirius in 1844, Bessel was led to believe that the brilliant orb was accompanied by another body, whose gravitational attraction was responsible for the irregularities observed in the path of the great dog-star when pursuing his journey through space. The elements of this hypothetical body were afterwards computed by Peters and Auwers, and its exact position assigned by Safford in 1861.

On January 31, 1862, Mr. Alvan Clarke, of Cambridgeport, Massachusetts, when engaged in testing a recently constructed telescope of great power, directed it on Sirius, and was enabled by good fortune to discover the companion star at a distance of ten seconds from its primary. Since its discovery, the star has pursued with such precision the theoretical path previously assigned to it that astronomers have had no hesitation in identi-

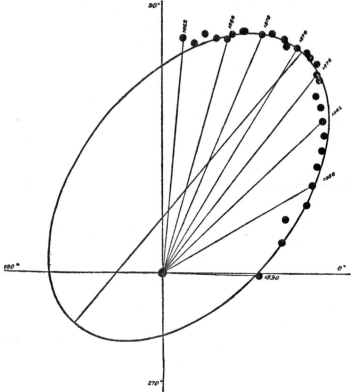

Fig. 5.—Apparent Orbit of the Companion of Sirius.
(*Drawn by Mr. Burnham.*)

fying it as the hypothetical body whose existence Bessel had correctly surmised.

The Sirian satellite is a yellow star of the eighth magnitude, and shines with a feeble light when contrasted with the surpassing brilliancy of its neighbour.

Astronomers were for some time in doubt as to whether the uneven motion which characterised the path of Sirius could be ascribed to the attraction of its obscure attendant, which presented such a marked contrast to its primary, and several observers were inclined to believe that the disturbing body still remained undiscovered. When, however, the density of the lesser star became known, it was discovered that, weight for weight, that of Sirius exceeded it only in the proportion of two to one, though as a light-giver the great orb is believed to be 5,000 times more luminous. The Sirian satellite revolves round its primary in about fifty years, and at a distance twenty-eight times that of the Earth from the Sun.

The surpassing brilliancy of Sirius as compared with that of the other stars of the firmament has rendered it at all times an object of interest to observers. The Egyptians worshipped the star as Sothis, and it was believed to be the abode of the soul of Isis. The nations inhabiting the region of the Nile commenced their year with the heliacal rising of Sirius, and its appearance was regarded as a sure forerunner of the rising of the great river, the fertilising flood of which was attributed to the influence of this beautiful star. It is believed that the Mazzaroth in Job is an allusion to this brilliant orb. Among the Romans Sirius was regarded as a star of evil omen; its appearance above the horizon after the summer solstice was believed to be associated with pestilence and fevers, consequent upon the oppressive heat of the season of the year.   The *dies caniculares*, or dog-days, were reckoned to begin twenty days before, and to continue for twenty days after, the heliacal rising of Sirius, the dog-star.   During those days a peculiar influence was believed to exist which created diseases in men and madness among dogs.   Homer alludes to the star

'whose burning breath
Taints the red air with fevers, plagues and death.'

Sirius, which is in Canis Major (one of Orion's hunting dogs), is a far more glorious orb than our Sun.   According to recent photometric measurements it emits seventy times the quantity of light, and is three times more massive than the great luminary of our system.   At the distance of Sirius (fifty billions of miles) the Sun would shrink to the dimensions of a third-magnitude star, and the light of seventy such stars would be required to equal in appearance the brilliant radiance of the great dog-star.   The orb, with his retinue of attendant worlds —some of which are reported as having been seen —is travelling through space with a velocity of not less than 1,000 miles a minute.

An irregularity of motion resembling that of Sirius has been detected with regard to Procyon, the lesser dog-star.   But in this case the companion star has not as yet been seen, though a careful search has been made for it with the most powerful of telescopes.   Should it be a planetary body, illumined by its primary, its reflected light would not appear visible to us, even if it were much less remote than it is.

We are able only to perceive the effulgence of brilliant suns scattered throughout the regions of space ; but besides those, there are doubtless many faintly luminous orbs and opaque bodies of vast dimensions occupying regions unknown to us, but by a knowledge of the existence of which an enlarged conception is conveyed to our minds of the greatness of the universe.

The most rapid of known revolving pairs is δ Equulei. The components are so close that only the finest instruments can separate them, and this they cannot do at all times. They accomplish a revolution in eleven and a half years. The slowest revolving pair is ζ Aquarii. The motion of the components is so tardy that to complete a circuit of their orbits they require a period of about sixteen centuries. Other binary stars have had different periods assigned to them; eleven pairs have been computed to revolve round each other in less than fifty years, and fifteen in less than 100 but more than fifty. There are other compound stars whose motions appear to be much more leisurely than those just mentioned, and although no orbital movement has, so far, been detected among them, yet, so vast is the scale upon which the sidereal system is constructed, that thousands of years must elapse before they can have accomplished a revolution of their orbits.

The Pole Star is an optical double, but the components are of very unequal magnitude. The

Pole Star itself is of the second magnitude, but its companion is only of the ninth, and on account of its minuteness is regarded as a good test for telescopes of small aperture. Mizar, in the constellation Ursa Major, is a beautiful double star. The components are wide apart, and can be easily observed with a small instrument.

There is a remarkable star in the constellation of the Lyre (ϵ Lyræ), described as a double double. This object can just be distinguished by a person with keen eyesight as consisting of two stars; when observed with a telescope they appear widely separated, and each star is seen to have a companion, the entire system forming two binary pairs in active revolution. The pair which first cross the meridian complete a revolution in about 2,000 years; the second pair have a more rapid motion, and accomplish it in half that time. The two pairs are believed to be physically connected, and revolve round their common centre of gravity in a period of time not much under one million years.

Cor Caroli, in Canes Venatici, is a pleasing double star, the components being of a pale white and lilac colour.

Albireo, in the constellation of the Swan, is one of the loveliest of double stars. The larger component is of the third magnitude, and of a golden yellow colour; the smaller of the sixth magnitude, and of a sapphire blue.

ϵ Bootis, known also as Mirac, and called by

Admiral Smyth 'Pulcherrima,' on account of its surpassing beauty, is a delicate object of charming appearance. The components of this lovely star are of the third and seventh magnitudes: the primary orange, the secondary sea-green.

The late Mr. R. A. Proctor, in describing a binary star system, writes as follows: ' If we regard a pair of stars as forming a double sun, round which—or, rather, round the common centre of which—other orbs revolve as planets, we are struck by the difference between such a scheme and our own solar system ; but we find the difference yet more surprising when we consider the possibility that in some such schemes each component sun may have its own distinct system of dependent worlds. In the former case the ordinary state of things would probably be such that both suns would be above the horizon at the same time, and then, probably, their distinctive peculiarities would only be recognisable when one chanced to pass over the disc of the other, as our Moon passes over the Sun's disc in eclipses. For short intervals of time, however, at rising or setting, one or other would be visible alone ; and the phenomena of sunset and sunrise must therefore be very varied, and also exquisitely beautiful, in worlds circling round such double suns. But when each sun has a separate system, even more remarkable relations must be presented. For each system of dependent worlds, besides its own proper sun, must have another sun—less splendid, perhaps (because farther off), but still brighter beyond comparison than our moon at the full. And, according to the position of any planet of either system, there will result for the time being either an interchange of suns, instead of the change from night to day, or else double sunlight during the day, and a corresponding intensified contrast between night and day. Where the two suns are very unequal or very differently coloured, or where the orbital path of each is very eccentric, so that they are sometimes close together and at others far apart, the varieties in the worlds circling round either, or around the common centre of both, must be yet more remarkable. " It must be confessed," we may well say with Sir John Herschel, " that we have here a strangely wide and novel field for speculative excursions, and one which it is not easy to avoid luxuriating in." '

Anyone who takes a cursory glance at the heavens on a clear night can readily perceive that there exists considerable diversity of colour among the stars. The contrast between some is pronounced and well marked, whilst others exhibit refined gradations of hue.

The most numerous class of stars are those which are described as white or colourless. They comprise about one-half of the stars visible to the naked eye. Among the most conspicuous examples of this type are Sirius—whose diamond blaze is sometimes mingled with an occasional flash of blue and red—Altair, Spica, Castor, Regulus, Rigel, all the stars of Ursa Major with the exception of one, and Vega

—a glittering gem of pale sapphire, almost colourless. The light emitted by stars of this class gives a continuous spectrum, the predominating element being hydrogen, having a very elevated temperature and under relatively high pressure. The vapours of iron, sodium, magnesium, and other metals, are indicated as existing in small quantities.

The second class of stars is that to which our Sun belongs. They are of a yellow colour, and embrace two-thirds of the remaining stars. The most prominent examples of this type are Arcturus, Capella, Aldebaran, Procyon, and Pollux. Hydrogen does not predominate so much in these as in the Sirian stars, and their spectra resemble closely the solar spectrum, indicating that they are composed of elements similar to those which exist in the Sun.

The star which bears the nearest resemblance to our Sun, both as regards the colour of its light and physical structure, is Capella, the most conspicuous star in the constellation Auriga, and one of the leading brilliants in the Northern Hemisphere. Its spectrum presents all the characteristics observed in the solar spectrum, and there exists an almost identical similarity in their physical constitution, though Capella is a much more magnificent orb than the Sun.

The third class of stars includes those which are of a ruddy hue, such as Betelgeux in the right shoulder of Orion, Antares in Scorpio, and a Herculis. Their spectra present a banded or columnar appearance, and there is greater absorption, especially of the blue rays of light. It is believed that the temperature of stars of this colour is not so elevated as that of those belonging to the other two orders, and that this is a sufficient reason to account for the different appearance of their spectra.

The aid of a good telescope is, however, necessary to enable us to perceive the varied colours and tints of the sparkling gems with which Nature has adorned her star-built edifice of the universe. Most of the precious stones on Earth have their counterparts in the heavens, presenting in a jewelled form contrasts of colour, pleasing harmonies, and endless variety of shade. The diamond, sapphire, emerald, amethyst, topaz, and ruby sparkle among crowds of stars of more sombre hue. Agate, chalcedony, onyx, opal, beryl, lapis-lazuli, and aquamarine are represented by the radiant sheen emanating from distant suns, displaying an inexhaustible variety of colour, blended in tints of untold harmony.

It is among double stars that the richest and most varied colours predominate. There are pairs of white, yellow, orange, and red stars; yellow and blue, yellow and pale emerald, yellow and rose red, yellow and fawn, green and gold, azure and crimson, golden and azure, orange and emerald, orange and lilac, orange and purple, orange and green, white and blue, white and lilac, lilac and dark purple, &c., &c. There are companion stars revolving round their primaries, coloured olive, lilac, russet, fawn, dun, buff, grey, and other shades indistinguishable by any name.

Our knowledge of binary star systems brings us to what may be regarded as the threshold of the fabric of the heavens. For it is known that other systems exist into the construction of which numerous stars enter. These form intricate and complex stellar arrangements, in which the component stars are physically united and retained in their orbits by their mutual attraction.

CHAPTER VII

THE STARRY HEAVENS

TRIPLE, QUADRUPLE, AND MULTIPLE STARS.—
These, when observed with the naked eye, appear
as single stars, but, when examined with a high
magnifying power, each lucid point can be resolved
into several component stars. They vary in number
from three to half a dozen or more, and form systems
of a more complex character than what are observed
in the case of binary stars.   In the usual construction of a triple system, the secondary star of a
binary is resolvable into two, each star being in
mutual revolution, whilst they both gravitate round
their primary.   By another arrangement, a close
pair control the movements of a distant attendant.

One of the most interesting of triple stars is the
tricoloured γ Andromedæ. The brilliant components of this system have their counterparts in the
topaz, the emerald, and the sapphire—the larger
star is of the third magnitude and of a golden yellow
colour ; the secondary of the fifth magnitude and of
an emerald green. These stars are ten seconds
apart, and, though they have been under observation since 1777, no orbital movement has as yet been
detected, but their common proper motion indicates

their close relationship and physical connection. In 1842, Otto Struve discovered that the companion star is itself double, and round it there gravitates a sapphire sun, which is believed to accomplish a revolution of its orbit in about 500 years. If round those suns there should be circling planetary systems of worlds inhabited by intelligent beings, the varied effects produced by the light emanating from those different coloured orbs would be of a very beautiful and pleasing nature.

A system suggestive of the endless variety of stellar arrangement that exists throughout the sidereal regions is apparent in the case of the triple star ζ Cancri. Two of the stars, of magnitudes six and seven, form a binary in rapid revolution, the components of which complete a circuit of their orbits in fifty-eight years, whilst the more distant third star, of almost similar magnitude, accomplishes a wide orbital ellipse round the other two in 500 or 600 years. These stars have been closely observed by astronomers during the past forty years, with the result that their motions have appeared most perplexing, and complicated beyond precedent. 'If this be really a ternary system,' wrote Sir John Herschel, 'connected by the mutual attraction of its parts, its perturbations will present one of the most intricate problems in physical astronomy.' The second star revolves round its primary, whilst the third pursues a retrograde course, but its path, instead of being even, presents the appearance of a series of circular loopings, in traversing which the star alternately quickens and slackens its pace, or at times appears to be stationary.

Astronomers have arrived at the conclusion that these perturbations are produced by the presence of a fourth member, which, though invisible, is probably the most massive of the system—perhaps a magnificent world teeming with animated beings, and attended by three suns which gravitate round it, dispensing light and heat to meet the requirements of the various forms of life which exist on its surface. In this system we have an arrangement the reverse of what exists in the solar system, where all the planets revolve round a predominant sun; but here there is a strange verification of the old Ptolemaic belief with regard to the path of a sun, though in this instance there are three suns circling round a dark globe which they illumine and vivify.

Triple stars occur with comparative frequency throughout the heavens. In Monoceros there is a fine triple star, discovered by Herschel, which he describes as 'one of the most beautiful sights in the heavens.' The stars ξ and β Scorpii form triple systems in which the components are differently arranged. In ξ the primary and secondary consist of two revolving stars which control the movements of a distant attendant; in β the primary and secondary stars are in mutual revolution, whilst round the former there circles a very close minute companion. There are doubtless many binary stars which, if examined with adequate telescopic power, would resolve themselves into triple and multiple systems, but the profound distances of those objects render the detection of their components a most difficult task.

Quadruple stars are usually arranged in pairs, *i.e.* the primary and secondary of a binary system are each resolvable into two, forming two pairs, each pair being in mutual revolution, while they both gravitate round their common centre of gravity. ε Lyræ, which has been described as a double double, is an example of a quadruple system, and ν Scorpii is of a similar construction, but more beautiful because its components are in closer proximity to each other. Close upon twenty of those double double systems have been discovered in different parts of the heavens.

One of the most interesting of quadruple systems is θ Orionis, which is situated in the Great Nebula, by which it is surrounded. This star, when observed with a telescope of low power, can be at once resolved into four separate lucent points, so arranged as to form a quadrilateral figure or trapezium. They are of the fifth, sixth, seventh, and eighth magnitudes, and described as pale white, garnet, faint lilac, and red. Though they have been under careful observation for upwards of two centuries, no perceptible motion has been perceived as occurring among them, nor has there been any change in their relative positions—they appear to be perfectly motionless; but we must not infer from this that no physical bond of union exists between them, for they are situated at an amazing distance from the Earth. Ascending higher in the scale of celestial architecture, we have multiple stars forming systems still more elaborate and complex, into the structure of which numerous stars enter, and they, as they increase in number, gradually merge into star-clusters.

If we assume that around each of the components of a multiple star there circles a retinue of planetary worlds, we are confronted with a most perplexing problem as to how the dynamical stability of a system so different from, and so vastly more complicated than, that of our solar system is maintained —where, as it were, suns and planets intermingle— how numerous circling orbs can accomplish their revolutions without being swayed and deflected from their paths by the gravitational attraction of adjacent members of the same system. Perplexing though the arrangement of such a scheme may be to our conception, yet, each orb has been weighed, poised, and adjusted by Infinite Wisdom, to perform its intricate motions in synchronous harmony with other members of the system—all moving in unison like the parts of a complicated piece of mechanism, and maintained in stable equilibrium by their mutual attraction—

> Mystical dance, which yonder starry sphere
> Of planets and of fixed in all her wheels
> Resembles nearest ; mazes intricate,
> Eccentric, intervolved, yet regular
> Then most, when most irregular they seem ;
> And in their motions harmony divine
> So smooths her charming tones that God's own ear
> Listens delighted.—v. 620-27.

All the natural phenomena with which we are familiar would, in the case of planets revolving round the component suns of a multiple system, be of a different kind or altogether absent. Instead of being illumined by one sun, those worlds would, at certain times, have several suns—some more distant than others—above their horizons, and upon very rare occasions, if ever, would there be an entire absence of all of those orbs from their skies. Consequently there would be no year such as we are familiar with ; no regular sequence of seasons similar to what is experienced on Earth ; no alternation of day and night, for there would be ' *no night there*,' though, in the absence of the primary orb, the light emitted by distant suns, whilst sufficient to banish night, and beyond comparison brighter than the Moon when at full, would, in the diminution of its intensity from that of noonday, be as grateful a change as that of from day to night which occurs on our globe.

Should those suns be differently coloured, each emitting its own peculiar shade of light as it appears above the horizon, the varied aspects of the perpetual day enjoyed by the inhabitants of those circling worlds present to the imagination harmonies of light and shade over which it is pleasant to linger.

TEMPORARY, PERIODICAL, AND VARIABLE STARS.— It may seem remarkable that among so many thousands of stars which spangle the firmament, there should occur no very perceptible change or variation in their aspect and brilliancy. From age to age they present the same appearance, shine with the same undiminished splendour, and rise and set with the same regularity. So that from time immemorial the stars have been regarded by mankind as the embodiment of all that is eternal and unchangeable. Yet, the serenity of the celestial regions does not always remain undisturbed—at occasional times a ' Nova,' or new star, blazes forth unexpectedly in the heavens, and perplexes astronomers ; and, after shining with a varying degree of brilliancy for a few weeks or months, gradually diminishes in size and brightness and eventually becomes lost to sight.

A record has been kept of about twenty temporary stars that have been observed at various periods since the time that reliable data of those objects have been published. Pliny mentions the appearance of a new star in the time of Hipparchus (134 B.C.) ; it was seen in the constellation of the Scorpion, and it is said that it was the apparition of this star which induced the celebrated astronomer to construct what is known as the earliest star catalogue. A new star is said to have become visible when the Emperor Honorius ruled, and another during the reign of the Emperor Otho, about 945 A.D. In May 1012 a new star appeared in Aries, and in July 1203 another was observed in Scorpio, which resembled Saturn. The most remarkable star of this kind was one observed by Tycho Brahé, which appeared in the constellation

Cassiopeia. He first perceived it on November 11, 1572. In lustre it equalled Jupiter, and when at its brightest rivalled Venus ; it was visible at noonday, and at night its light could be perceived through strata of cloud which rendered all other stars invisible. The star maintained its brilliancy for three weeks, when it became of a yellowish colour and perceptibly decreased in size ; it afterwards assumed a ruddy hue resembling Aldebaran, and, diminishing gradually in magnitude and brightness, ceased to be visible in March 1574. It twinkled more than the other stars, and during the time it could be perceived its position remained unchanged. In 1604 a conspicuous new star burst forth in Ophiuchus. It surpassed in brilliancy stars of the first magnitude, and outshone the planet Jupiter, which was in its proximity. Kepler observed this star, and described it as ' sparkling like a diamond with prismatic tints.' It soon began to decline after its appearance ; in March 1605 it had shrunk to the dimensions of a third-magnitude star, and in a year later it became entirely lost to view. Other stars of the same class, though of a less conspicuous character, have been observed at occasional times. Anthelme, a Carthusian monk, discovered one near β Cygni in 1670 ; another appeared in Ophiuchus in 1848 ; one in Scorpio in 1860 ; one in Corona Borealis in 1866 ; in Cygnus in 1876 ; in Andromeda in 1885 ; and in Auriga in 1892.

Various theories have been advanced in order to account for the sudden outbursts of those stars, the light from which has probably occupied not much less than one hundred years in its passage hither. It has been suggested that the collision of two suns, or of two great masses of matter, would create such phenomena ; but, apart from the improbability of such a catastrophe occurring among the celestial orbs, the rapid subsidence in the luminosity of the observed objects would indicate that the outburst was produced by causes of a more rapidly transitory nature than what would result from the collision of two condensed masses of matter. A collision occurring between two swarms of meteors has been suggested as one way of accounting for the sudden appearance of those stars ; but another, and more plausible, explanation is that they are produced by a great eruption of glowing gas from the interior of a sun, causing an enormous increase in its luminosity, which subsides after a time, and is succeeded by a normal condition of things. It has been observed that all those temporary stars, with the exception of two, have appeared in the region of the Milky Way. In this luminous zone the condensation of small gaseous stars and nebulæ is more pronounced than in any other part of the heavens, and this would seem to indicate that there may be cosmical changes taking place among them which need not be associated with the occurrence of catastrophes resulting in the conflagration of worlds, and that Nature, in accomplishing her purposes, does not overstep the uniform working of her laws, upon which depend the stability and existence of the universe.

PERIODICAL AND VARIABLE STARS are distinguished from other similar objects by the fluctuations which occur in the quantity of light emitted by them. The difference in the luminosity of some stars is at times so marked that, in a few weeks or months, they decline from the first or second magnitudes to invisibility, and, after the expiration of a certain period, they again gradually regain their pristine condition. When these changes take place with regular recurrence, they are called 'periodical;' when they occur in a variable and uncertain manner, they are called 'irregular.' About 300 stars are known as variable, but the majority of them are telescopic objects. Their periodical changes of brilliancy present every degree of variety; in some stars they are scarcely perceptible and occur at long intervals; in others, changes of brightness occur in a few hours or days, by which the light emitted is intensified many hundreds of times.

Some stars accomplish their cycle of change in a few days, many in a few weeks or months, and there are others which do not complete their periods until the expiration of a number of years.

One of the most remarkable of variable stars is called Mira 'the wonderful,' in the constellation Cetus. When at its maximum brilliancy it shines for two or three weeks as a star of the second magnitude. It then begins to gradually decline, and at the end of three months becomes invisible. It remains invisible for five months, and then reappears, and during the ensuing three months it regains by

P

degrees its former brilliancy. Mira completes a
cycle of its changes in 334 days, and, during that
time, oscillates between a star of the second and
tenth magnitude. The variability of Mira Ceti was
first observed by David Fabricius in the sixteenth
century.

Another remarkable star is η Argus, which is surrounded by the great nebula in the constellation Argo
Navis. It is invisible to the naked eye, but in the
telescope it has a reddish appearance, and is slightly
brighter than the stars in its vicinity. It was first
observed by Halley in 1677, and it was then of the
fourth magnitude. In 1751 it had risen to the
second magnitude, and maintained its position as a
star of this class until 1837, when, on December 16
of that year, its brilliancy suddenly increased, and it
equalled in a short time α Centauri. It reached its
maximum in 1843, and then it was surpassed only
by Sirius. It maintained its brilliancy for about
ten years. In 1858, it declined to the second magnitude, in 1859 to the third, and, gradually diminishing, it became invisible to the naked eye in 1868.
It is now of the seventh magnitude, and is again
increasing, and may soon resume its position among
the other stars. It is believed to have a period of
seventy years, and in that time its light ebbs and
flows between the seventh and first magnitudes.

The most interesting variable star in the heavens
is Algol (the demon), in the constellation Perseus.
Its light fluctuations can be observed without the
aid of a telescope, and it completes a cycle of its
changes in two or three days. For about two days

and thirteen hours it is conspicuously visible as a
star of the second magnitude ; it then begins to
decline, and in about four hours sinks to the dimensions of a fourth-magnitude star ; it remains in this
condition for twenty minutes, and then increases
gradually until, at the expiration of four hours, it
regains its former brilliancy, which it sustains for
two days and thirteen hours, when it again goes
through the same cycle of changes in a precisely
similar manner to what has been described. Astrologers have ascribed many evil influences to the
demon star, which adorned the head of Medusa ;
nor did it escape the observation of ancient astronomers that this malevolent orb is—as a modern
writer amusingly remarks—slowly winking at us
from out the depths of space.

Variable stars are found in greater numbers in
some parts of the heavens than in others. Those
of a white colour, and with shorter and more regular
periods, are most numerous in the region of the
Milky Way ; those that are small, with long periods
and of a reddish hue, are more widely removed from
that zone. Stars of this class are all very remote,
and no attempt has as yet been made to ascertain
the parallax of Algol.

Several theories have been suggested in order to
account for the periodical brilliancy of those stars.
It has been suggested that the stars have opaque
non-luminous patches on their surfaces, and that
during axial rotation their light ebbs and flows according as the dark or bright portions are turned

towards us. This theory is highly improbable. Another and more plausible reason, especially with regard to short period variables, is, that around those stars there revolve opaque bodies or satellites which at times intercept a portion of their light by producing a partial eclipse of their discs, similar to that caused by the dark body of the Moon when passing between the Sun and the Earth.

It is now known that in the case of variables of the Algol type, the periodical fluctuations of their light arises from this cause, and that round Algol there is a dark world or satellite travelling, which completes a revolution of its orbit in about sixty-nine hours, and that, during each circuit, it intercepts one half of the light of its primary by partially eclipsing the orb, and thereby creating a diminution in its apparent magnitude which becomes perceptible at recurring intervals.

STAR GROUPS.—These are plentifully scattered over the heavens and, by their conspicuous brilliancy, add to the grandeur and magnificence of the midnight sky. The Hyades in Taurus, of which Aldebaran is the chief, forming the eye of the Bull, attract attention.

The stars in Coma Bernices form a rich group ; the sickle in Leo, the seven stars in Ursa Major, and those in Cassiopeia and Aquila are familiarly known to all observers. Besides these, there are many other groups and aggregations of stars which adorn the celestial vault and enhance the beauty of the heavens.

STAR CLUSTERS.—On observing the heavens on a clear, dark night, there can be seen in different parts of the sky closely aggregated groups of stars called clusters. In some instances the component stars are so near together that the naked eye is unable to discern the individual members of the cluster. They then assume an indistinct, hazy, cloudlike appearance. Upwards of 500 clusters are known to astronomers, the majority of which are very remote. Many of them contain thousands of stars compressed into a very small space, and others are so distant that the largest telescopes are incapable of resolving their nebulous appearance into separate stars.

Star clusters have been arranged into two classes, 'irregular' and 'globular;' but no sharp line of demarcation exists between them, though each have their distinctive peculiarities. Irregular clusters consist of aggregations of stars brought promiscuously together, and presenting an appearance devoid of any structural arrangement. They are of different shapes and sizes, possess no distinct outline, and are not condensed towards their centre, like those that are globular. On examination, they present an intricate reticulated appearance; streams and branches of stars extend outwards from the parent cluster, sometimes in rows and sinuous lines, and, in other instances, diverging from a common centre, forming sprays. Sometimes the stars are seen to follow each other on the same curve which terminates in loops and arches of symmetrical proportions.

There are three conspicuous clusters in the northern sky that are visible to the naked eye—viz. the Pleiades in Taurus, the Great Cluster in the sword-handle of Perseus, and Praesepe in Cancer, commonly called the Beehive.

The cluster which from time immemorial has had bestowed upon it the chief attention of mankind are the beautiful Pleiades or Seven Sisters, and intertwined among its stars are the legendary and mythological beliefs of ancient nations and untutored tribes inhabiting the different regions of the globe. When viewed with a telescope of moderate size the cluster appears as a scattered group, and numerous stars become visible that are imperceptible to ordinary vision.

In the sword-handle of Perseus there is a cluster which, to the naked eye, appears as a small patch of luminous cloud. This inconspicuous object when observed with an instrument of moderate power is resolved into a magnificent assemblage of stars, and presents a spectacle which creates in the mind of the beholder mingled feelings of admiration and amazement. No telescope has yet penetrated its utmost depths, or revealed all the glories of this shining region, crowded with glittering points of light comparable in number to the pebbles strewn on the shore of a troubled sea.

The cluster Praesepe in Cancer is visible on a clear night to the unaided eye as a small nebula. This object attracted the attention of Galileo, to which he applied his newly invented telescope, and was delighted to find that his glass was capable of resolving it into a group of stars thirty-six in number, and all of comparatively large magnitude. The disappearance of Praesepe in consequence of the condensation of vapour in the atmosphere was regarded by the ancients as a sure indication of approaching rain. In the same constellation, near the Crab's southern claw, there is another rich cluster, which consists of 200 stars of the ninth and tenth magnitudes.

In Sobieski's Shield there is a magnificent fan-shaped cluster of minute stars with a prominent one in its centre; and in the constellation of the Southern Cross there is a cluster which, on account of the varied colours of its component stars, has been compared by Sir John Herschel to 'a piece of rich fancy jewellery;' eight of the principal stars being coloured red, green, and blue.

GLOBULAR CLUSTERS.—These have been described by Herschel as 'the most magnificent objects that can be seen in the heavens.' They are all very remote, of a rounded form, and when viewed with a telescope present the appearance of 'a ball of stars.' In some clusters the constituent stars are distinguishable as minute points of light; in others, more remote, they are of a coarse granular texture, and in those still more distant they resemble a 'heap of golden sand.' Some clusters are situated at such a profound distance in space that it is impossible with the most powerful of telescopes to define their stellar structure; all that can be distinguished of these is a cloudy luminosity resembling in appearance an irresolvable nebula. Globular clusters usually present a radiated appearance. Rays, branches, and spiral-shaped streams of stars appear to flow from the circumference of some ; and, in other instances, fantastic appendages of stars project outwards from the parent cluster. There doubtless exists much variety in the structural arrangement of these clusters, and an equal diversity in the magnitude and number of the stars which enter into their formation. The stars in some clusters may equal those of the first magnitude, and in others they may not exceed in dimensions the minor planets. In the telescope they vary in size from the eleventh to the fifteenth magnitude ; the smaller stars occupy the centre of a cluster, whilst the larger ones are found near its circumference. Globular clusters are more condensed towards their centre than those of irregular shape, and some have a nucleated appearance. This apparent condensation is not altogether owing to the depth of star strata as viewed from the circumference of the cluster, but there appears to exist an attractive force (probably gravitational) which draws the stars towards its centre, and if this 'clustering power' were not opposed by some other counteracting force, those bodies would coalesce into one mass. It may be 'that a centrifugal impulse predominates by which full-grown orbs are driven from the nursery of suns in which they were reared to seek their separate fortunes and enter on an independent career elsewhere.'

It is not known how the dynamical equilibrium of a star cluster is maintained ; and on account of its extreme distance no motion is perceptible among its component stars. The laws by which those stellar aggregations are produced and governed are wrapped in obscurity, and the nature of the motions of their stars, whether towards concentration or diffusion, cannot at present be ascertained. If those globular clusters could be observed sufficiently near, they would most probably expand into vast systems of suns occupying immense regions of space.

The largest and most magnificent globular cluster in the heavens is ω Centauri, in the Southern Hemisphere. To the naked eye it resembles a round, indistinct, cometary object, about equal to a star of the fourth magnitude ; but when observed with a powerful telescope it appears as a globe of considerable dimensions composed of innumerable stars of the thirteenth and fifteenth magnitudes, all exceedingly minute and gathered into small knots and groups. A remarkable cluster in Toucani is described by Sir John Herschel as ' most magnificent ; very large ; very bright, and very much compressed in the middle.' The interior mass consists of closely aggregated pale rose-coloured stars, surrounded by others of a pure white which embrace the remainder of the cluster. There is a fine globular cluster in Sagittarius between the Archer's head and the bow. It was observed by Hevelius in 1665. The central portion is very much compressed, and consists of excessively minute stars enclosed by others of larger size. In Aquarius there is a magnificent ball of stars of a beautiful spherical form, which Sir J. Herschel compared to a heap of fine sand. Numerous other clusters are profusely distributed over the heavens, occupying regions in the profound depths of space which can only be reached by the aid of most powerful instruments.

The finest and most remarkable object of this class visible in the northern heavens is the Great Cluster which lies between η and ζ Herculis. It was discovered by Halley in 1714, who writes : ' This is but a little patch, but it shows itself to the naked eye when the sky is serene and the moon absent.' When observed with a powerful telescope its magnificence at once becomes apparent to the beholder. ' Perhaps,' says Dr. Nichol, ' no one ever saw it for the first time through a telescope without uttering a shout of wonder.' At its circumference the stars are rather scattered, but towards the centre they appear so closely aggregated that their combined effulgence forms a perfect blaze of light. Sir William Herschel estimated that there are 14,000 stars in the cluster, each a magnificent world but unaccompanied by any planetary attendants.

As a result of more recent investigations this number has been considerably reduced, and it is now generally believed that about 4,000 stars enter into the formation of the cluster. As its distance from the Earth is unknown, it follows that

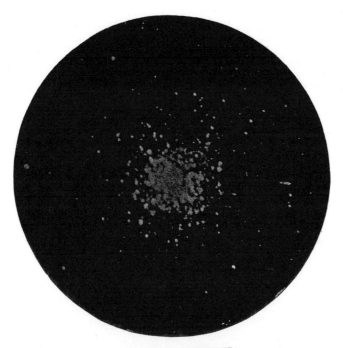

CLUSTER IN HERCULES

there must be some uncertainty attached to any conclusions that may be arrived at with regard to this superb object. Miss Agnes Clerke estimates the number of the constituent stars at 4,000, and in support of her conclusion this talented lady writes as follows: 'The apparent diameter of this object, including most of the "scattered stars in streaky masses and lines" which form a sort of " glory " round it, is 8'; that of its truly spherical portion may be put at 5'. Now, a globe subtending an angle of 5' must have (because the sine of that angle is to radius nearly as to 1 : 687) a real diameter $\frac{1}{687}$ of its distance from the eye, which, if we assume to be such as would correspond to a parallax of $\frac{1}{20}$ of a second, we find that the cluster, outliers apart, measures 558,000 millions of miles across. Light, in other words, occupies thirty-six days in traversing it, but sixty-five years in journeying thence hither. Its components may be regarded, on an average, as of the twelfth magnitude; for, although the divergent stars rank much higher in the scale of brightness, the central ones, there is reason to believe, are notably fainter. The sum total of their light, if concentrated into one stellar point, would at any rate very little (if at all) exceed that of a third-magnitude star. And one star of the third is equivalent to just four thousand stars of the twelfth magnitude. Hence we arrive at the conclusion that the stars in the Hercules Cluster number much more nearly four than fourteen thousand.'

For what purpose do those thousands of clustering orbs shine? Who can tell? Night is unknown
in the regions illumined by their brilliant radiance.
This stupendous aggregation of suns testifies to the
magnificence of the starry heavens, and to the
omnipotence of the Creator.

GALAXIES.—These consist of vast aggregations
of stars which form separate 'island universes'
floating in the depths of space; they are believed
to equal in magnitude and magnificence the Milky
Way—the galaxy to which our system belongs.

NEBULÆ.—We now reach the last, and what are
believed to be the most distant of the known con-
tents of the heavens. They are all exceedingly
remote, devoid of any perceptible motion, faintly
luminous, and, with the exception of two of their
number, invisible to the naked eye. Halley was the
first astronomer who paid any attention to those
objects. In 1716 he enumerated six of them, but
of this number only two can, in a strict sense, be regarded as nebulæ, the others since then have been
resolved into magnificent star clusters. In 1784,
Messier catalogued 103 nebulæ, and the Herschels—
father and son—in their survey of the stellar regions,
discovered 4,000 of those objects. There are now
8,000 known nebulæ in the heavens, but the majority
of them are not of much interest to astronomers.
Prior to the invention of the spectroscope it was
believed that all nebulæ were irresolvable star
clusters, but the analysis of their light by this instrument indicated that their composition was not

stellar but gaseous. Their spectra consist of a few bright lines revealing the presence of hydrogen, nitrogen, and other gaseous elements.

Much that is mysterious and uncertain is associated with those objects which appear to lie far beyond the limits of our sidereal system. It is now generally believed that they exhibit the earliest stage in the formation of stars and planets—inchoate worlds in process of slow evolution, which will eventually condense into systems of suns, and planetary worlds.

Nebulæ present every variety of form. Some are annular, elliptic, circular, and spiral; others are fan-shaped, cylindrical, and irregular, with tufted appendages, rays, and filaments. A fancied resemblance to different animated creatures has been observed in some. In Taurus there is a nebula called the ' Crab ' on account of its likeness to the crustacean ; another is called the ' Owl Nebula ' from its resemblance to the face of that bird. The Orion Nebula suggests the opened jaws of a fish or sea monster, hence called the Fish-Mouth Nebula. There is a Horse-Shoe Nebula, a Dumb-Bell Nebula, and many others of various shapes and forms. They are classified as follows : (1) Annular Nebulæ, (2) Elliptic Nebulæ, (3) Spiral Nebulæ, (4) Planetary Nebulæ, (5) Nebulous Stars, (6) Large Irregular Nebulæ.

ANNULAR NEBULÆ.—These resemble in appearance an oval-shaped luminous ring ; they are comparatively few in number, and not more than a dozen have been discovered in the whole heavens. The most remarkable object of this class is the Ring Nebula, which is situated between the stars β and γ Lyræ.   It is visible in a moderate-sized telescope as a well-defined, flat, oval ring ; its central part is not quite dark but is occupied by a filmy haze of luminous matter which is prolonged inwards from the margin of the ring.   When examined with a high power the edges of the ring have a fringed appearance, and numerous glittering stellar points become visible both within and without its circumference.   This nebulous ring, though a small object in the telescope, is of enormous magnitude, and if it were not more distant than 61 Cygni, one of the nearest of the fixed stars, its diameter would not be less than 20,000 millions of miles, but it has been estimated by Herschel that it is 900 times more remote than Sirius.   How stupendous, then, must be its dimensions, and how bewildering to our conception is the profound immensity of space in which it is located !   An annular nebula similar to that of Lyra, but on a smaller scale, is found in Cygnus, and within it there can be seen a conspicuous star. Another exists in Scorpio which contains two stars situated within the ring at diametrically opposite points to each other.

ELLIPTICAL NEBULÆ.—The most interesting object of this class is the Great Nebula in Andromeda, called ' the transcendentally beautiful queen of the nebulæ '—an appellation which it scarcely merits.   This object, which is plainly visible to the naked eye, is of an oval shape, of a milky white colour, and is situated near the most northern star of the three which form the girdle of Andromeda. It was known to the ancients, and Ali Sufi, a Persian astronomer who flourished in the tenth century, alludes to it ; but it did not attract much attention until the seventeenth century. Simon Marius was the first to observe this object with a telescope. This he did on December 15, 1612 ; he describes it as shining with a pale white light resembling in appearance the flame of a candle when seen through a semi-transparent piece of horn. When examined with a high magnifying power it is seen to occupy a largely extended area measuring 4° in length and $2\frac{1}{2}$° in breadth. Its luminosity increases from the circumference to the centre, where there can be seen a small nucleus with an ill-defined boundary, which has the appearance of being granular, but its composition is not stellar. Two dark channels running almost parallel to each other and to the axis of the nebula have been observed by Bond ; these, when prolonged, form into curves which terminate in two great rings. They are wide rifts which separate streams of nebulous matter, and are indicative that some formative processes may be going on within the nebula.

Astronomers have been baffled in their attempts to discover the nature of the Andromeda Nebula. Though great telescopes have been able to render visible thousands of stars over and around it, yet the nebula itself is irresolvable and bears no trace of stellar formation; neither, according to Dr. Huggins, is its spectrum gaseous, a circumstance which deepens the mystery associated with this object. Its distance is unknown, and its dimensions cannot be ascertained.

Other elliptical nebulæ are found in different regions of the heavens. In Ursa Major there is an oval nebula resembling that of Andromeda, but on a much smaller scale. It possesses a nucleus, and on the photographic plate there can be detected the presence of spiral structure, indicating the existence of streams of nebulous matter. Adjacent to this nebula is another of the same class with a double nucleus, and associated with it is a nebulous star.

SPIRAL NEBULÆ.—The great reflector of Earl Rosse at Parsonstown was the successful means by which nebulæ of this form were discovered. This powerful telescope was capable of defining with greater accuracy the structural formation of those objects than any other instrument in use. It was ascertained that spiral coils and convoluted whorls enter into the structure of most nebulæ, indicating a similarity in the process of change which may be going on in these vast accumulations of cosmical matter. The most interesting specimen of a spiral nebula is situated in Canes Venatici. It consists of spiral coils emanating from a centre with a nucleus and surrounded by a narrow luminous ring. In appearance it resembles the coiled mainspring of a watch.

PLANETARY NEBULÆ.—These have been so named

on account of the resemblance which they bear to the discs of planets. They are of uniform brightness, circular in shape, with sharply-defined edges, and are frequently of a bluish colour. They are more numerous than annular nebulæ; three-fourths of their number are in the Southern Hemisphere, and they are situated in or very near the Milky Way. Those objects were first described by Sir William Herschel, who was rather perplexed as to what was their real nature and how he should classify them. He remarked that they could not be planets belonging to far-off suns, nor distant comets, nor distended stars. Consequently, he concluded rightly that they were nebulæ. When observed with large telescopes, they lose their planetary aspect, and their sharpness of outline is less apparent; their discs become broken up into bright and dark portions, and in some, numerous minute stars have been observed, whilst others have well-defined nuclei.

The most prominent nebula of this class is situated in the constellation Ursa Major, and is called the Owl Nebula, from its fancied resemblance to the face of that bird. Sir John Herschel describes it as ' a most extraordinary object, a large, uniform nebulous disc, quite round, very bright, not sharply defined, but yet very suddenly fading away to darkness.' When examined in 1848 with Earl Rosse's reflector, two bright stars were discovered in its interior; each was in the centre of a circular dark space surrounded by whorls of nebulous

matter—hence the origin of its name.   This nebula
gives a bright line spectrum indicative of gaseous
composition.   It is believed to consist chiefly of
hydrogen and other gases which form a globe of
such stupendous magnitude that, if we surmise its
distance from the earth to be sixty-five light years—
an estimate much too low—'its diameter would
exceed that of the orbit of Neptune upwards of
100 times.'[1]   Within its compass the orbs of hundreds of solar systems as large as that of ours
would be able to perform their revolutions, having
spacious intervals existing between each system.
Another interesting planetary nebula is in the constellation of the Dragon, near to the pole of the
ecliptic ; it is slightly oval, of a pale blue colour,
and contains a star of the eleventh magnitude in
its centre.   It gives a gaseous spectrum.   Attempts
have been made to determine its parallax, but without success, and during the eighty years it has
been under observation it has remained apparently
motionless.   Its light period, if estimated at 140
years, would indicate the existence of a globe with a
diameter equal to forty-four diameters of the orbit
of the planet Neptune.[2]   A nebula of this class was
discovered by Sir John Herschel in the Centaur.
He described it as resembling Uranus, but larger ;
its colour was of a beautiful rich blue, and its light
equalled that of a star of the seventh magnitude.

NEBULOUS STARS.—These stars are each surrounded by a luminous haze several minutes of arc

[1] Miss Clerke's *System of the Stars.*   [2] *Ibid.*

in diameter and of a circular form. Sir William
Herschel, by his observation of those objects, arrived
at the conclusion 'that there exists in space a
shining fluid of a nature totally unknown to us,
and that the nebulosity about those stars was not
of a starry nature.' Thirteen stars of this type have
been enumerated by him and many others have
since been discovered. The 'glow' which surrounds them has been observed in a few instances
to have vanished without leaving any trace of
nebulosity behind, but the causes which have
brought about such a result are entirely unknown.
The nature of those stars is involved in considerable obscurity, and one class of nebula would seem
to merge into the other; nebulous stars with faint
aureolæ do not differ much from small nebulæ
interspersed with stellar points.

LARGE IRREGULAR NEBULÆ.—These are found in
both hemispheres, and are remarkable on account of
the varied appearances which they present, and the
large extent of space which many of them occupy.
In some, the nebulous matter of which they are composed can be seen like masses of tufted flocculi,
sometimes piled up, and at other times promiscuously
scattered, resembling in appearance the foam on the
crested billows of a surging ocean rendered suddenly
motionless, or cirro-cumuli floating in a tranquil sky.
Islands of light with intervening dark channels, pro-
montories projecting into gulfs of deep shade, sprays
of luminous matter, convoluted filaments, whorls,

wreaths, and spiral streams all enter into the structural formation of a great nebula.

The Great Nebula in Argo, in the Southern Hemisphere, is one of the most remarkable objects of this class. It consists of bright irregular masses of luminous matter, streaks and branches, and occupies an area about equal to one square degree. At its eastern border is situated the variable star η Argus, which fluctuates between the first and seventh magnitudes in a period of about seventy years.

A rich portion of the Galaxy lies in front of the nebula, which creates an effect as if it were studded over with stars. Sir John Herschel, in describing this nebula, writes as follows :—' The whole is situated in a very rich and brilliant part of the Milky Way, so thickly strewed with stars that, in the area occupied by the nebula, not less than 1,200 have been actually counted. Yet it is obvious that these have no connection whatever with the nebula, being, in fact, only a simple continuation over it of the general ground of the Galaxy. The conclusion can hardly be avoided that, in looking at it, we see through and beyond the Milky Way, far out into space, through a starless region, disconnecting it altogether from our system. It is not easy for language to convey a full impression of the beauty and sublimity of the spectacle which this nebula offers as it enters the field of view of a telescope, fixed in right ascension, by the diurnal motion, ushered in as it is by so glorious and innumerable a procession of stars, to which it forms a sort of climax, and in a part of the heavens otherwise full of interest.' Another large bright nebula (called 30 Doradus), also in the Southern Hemisphere, is composed of a series of loops with intricate windings forming a kind of open network against the background of the sky which it adorns. Sir John Herschel describes it as one of the most extraordinary objects in the heavens.

The 'Crab' Nebula in Taurus, the 'Horse-Shoe' Nebula in Sobieski's Shield, and the 'Dumb-Bell' Nebula in Vulpecula are remarkable objects, but the assistance of a powerful telescope is required to bring out their distinctive features. The 'Crab' Nebula is partially resolvable into stars ; the other two are believed to be gaseous.

The largest and most remarkable of all the nebulæ is that known as the Great Nebula in Orion, which was discovered and delineated by Huygens in the middle of the seventeenth century. It is perceptible to the naked eye, and when viewed with a glass of low power can be seen as a circular luminous haze surrounding the multiple star θ Orionis— one of the stars in the Giant's Sword, and which is of itself a remarkable object. The most conspicuous part of the nebula bears a slight resemblance to the wing of a bird ; it consists of flocculent masses of nebulous matter possessing a faint greenish tinge. Sir John Herschel compared it to a surface studded over with flocks of wool, or to the breaking up of a mackerel sky when the clouds of which it consists begin to assume a cirrous appearance. Its brightest portion is occupied by four conspicuous stars, which form a trapezium; around each there is a dark space free from nebulosity, a circumstance which would seem to indicate that the stars possess the power either of absorbing or of repelling the nebulous matter in their immediate vicinity. When observed with a powerful telescope, this nebula appears to be of vast dimensions, and, with its effluents, occupies an area of 4° by 5½°. Irregular branching masses, streams, sprays, filaments, and curved spiral wreaths project outward from the parent mass, and become gradually lost in the surrounding space. This object remained for long a profound mystery; no telescope was capable of resolving it, nor was it known what this 'unformed fiery mist, the chaotic material of future suns,' was, until the spectroscope revealed that it consists of a stupendous mass of incandescent gases—nitrogen, hydrogen, and other elementary substances, occupying a region of space believed by some to equal in extent the whole stellar system to which our Sun belongs.

In the Southern Hemisphere, near to the pole of the equator, are two nebulous clouds of unequal size; the larger having an area about four times that of the smaller. They are known as the Magellanic Clouds, having been called after the navigator Magellan. Both are visible on a moonless night, but in bright moonlight the smaller disappears. Sir John Herschel, when at the Cape of Good Hope, examined those objects with his powerful telescope. He described them 'as consisting of swarms of stars, globular clusters, and nebulæ of various kinds, some

GREAT NEBULA IN ORION

portions of them being quite irresolvable, and presenting the same milky appearance in the telescope that the nebulæ themselves do to the naked eye.' These are believed to be other universes of stars sunk in the profound depths of space, our knowledge of their existence being dependent upon the faint nebulous light which left them, perhaps, several thousand years ago.

The description of the various kinds of nebulæ leads us to consider what is called the Nebular Hypothesis. That the stars and solar system had at some time in the past a beginning, is as much a matter of certainty as that they will at some future time cease to be. Stars, like organic beings, have their birth, grow and arrive at maturity, then decline into a state of decrepitude, and finally die out. The duration of the life of a star, which may be reckoned by millions of years, depends upon the length of time during which it can maintain a temperature that renders it capable of emitting light. By the constant radiation of its heat into space, a condition of its constituent particles consequent upon the gradual contraction of its mass will ultimately occur, which will result in the exhaustion of its stores of thermal energy, the extinction of its light, and the reduction of what was once a brilliant orb to the condition of a mass of cold, opaque, inert matter. Inquiries as to the origin of the stars have led scientific men to conclude that they have been evolved from gaseous nebulæ, and these have therefore been regarded as indicating the earliest stage in the formation of suns and planets. It is believed that the condensation of those attenuated masses of luminous matter into stars is capable of accounting for the generation and formation of all the shining orbs which enter into the structure of the starry heavens. In the evolution of a 'cosmos out of a chaos' we should expect to find stars presenting every stage of development—some in an embryo state and others more advanced; stars in full vigour and activity, stars that have passed the meridian of life, and stars in a condition of decay and on the verge of extinction. The observations of astronomers have led them to conclude that this condition of 'youth and age' exists among the stellar multitude; but the characteristics by which it is distinguished are neither very obvious nor reliable.

The nebular theory is incapable of proof or demonstration; but modern discoveries tend to support the accuracy of its conclusions, and its principles have now been adopted by the majority of philosophic thinkers. The physical changes which are going on in the nebulæ towards stellar evolution, or in fully formed stars towards dissolution, are so slow that the life of an individual, or even the historical records of the past, are incapable of furnishing any evidence of alteration in their condition. A period of time infinitely greater than what has elapsed since the birth of science must pass before anything can be known of the life history of the stars; indeed, the allotted span of man's

existence on this planet may have terminated ere
the evolution of a large nebula into a star cluster
can have taken place.

The nebular hypothesis was first propounded by
Kant, who suggested that the sun and planets originated from a vast and diffused mass of cosmical
matter. This theory was afterwards supported
by Herschel and by the great French astronomer
Laplace. As a result of close and continued observation of the different classes of nebulæ, Herschel
arrived at the conclusion that there exists in space
a widely diffused 'shining fluid,' of a nature totally
unknown to us, and that the nebulosity which he
perceived to surround some stars was not of a starry
nature. He further adds that this self-luminous
matter 'seemed more fit to produce a star by its
condensation than to depend on the star for its existence.' His sagacious conclusion with regard to
the non-stellar nature of this nebulous matter was
afterwards confirmed by the spectroscope; for at
that time it was believed that even the faintest
nebulæ were irresolvable star clusters.

In 1811 Herschel read a paper before the Royal
Society in which he propounded his famous nebular
hypothesis, and stated his reasons for believing that
nebulæ, by their gradual condensation, were transformed into stars. Having assumed that there
exists a highly attenuated self-luminous substance
diffused over vast regions of space, he endeavoured
to show that by the law of attraction its particles
would have a tendency to coalesce and form aggre-

gations of nebulous matter, and that each of these,
by the continued action of the same force, would
gradually condense and ultimately acquire the con-
sistence of a star. In the case of large irregular
nebulæ, numerous centres of attraction would
originate in the mass, round which the nebulous
particles of matter would arrange themselves; each
nucleus, when condensation had been completed,
would become a star, and the entire nebula would
in this manner be transformed into a cluster of
stars. Herschel believed that he could trace the
different stages of nebular condensation which
result in the evolution of a star. In large, faintly
luminous nebulæ the process of condensation had
only commenced; in others that were smaller and
brighter it was in a more advanced stage; in
those that contained nuclei there was evidence of
nascent stars; and, finally, there could be seen in
some nebulæ minute stellar points—new-born suns
—interspersed among the haze of the transforming
mass. By this theory Herschel was able to account
for the phenomena associated with nebulous stars
and the supposed changes which were observed in
some nebulæ. The nebular hypothesis as described
by Herschel was not received with much favour,
nor did it unsettle much the belief that all nebulæ
were vast stellar aggregations, and that their cloudy
luminosity was a consequence of the inadequacy of
telescopic power to resolve them into their component stars. Laplace, who was highly gifted as a
geometrician, demonstrated how the solar system

could have been evolved in accordance with dynamical principles from a slowly rotating and slowly
contracting spheroidal nebula. The rotatory motion
of a nebula, in obedience to a well-known mechanical law, increases as its density becomes greater,
and this goes on until the tangential force at the
equator overcomes the gravitational attraction at
its centre. When this occurs, a revolving ring of
nebulous matter is thrown off from the parent mass,
and by this means equilibrium is restored between
the two forces. As the rotatory velocity of the
nebula continues to increase with its contraction,
another ring is cast off, and in this manner a succession of revolving rings may be detached from
the condensing spheroid ; each newly-formed ring
being nearer to the centre of the contracting mass
and revolving in a shorter period than its predeces-
sor. In the evolution of our system, the central
mass of the nebula became the Sun and each of
the revolving rings, by their condensation into one
mass, formed a planet. In a similar manner, though
on a diminished scale, the elementary planets,
whilst in a nebulous state, parted with annular
portions of their substance, out of which were
evolved their systems of satellites. This theory
furnished a plausible reason, which was capable of
explaining how the orbs which constitute the solar
system came into existence, and, though hypothetical, yet the manner in which it accounted for the
orderly and symmetrical genesis of the system
rendered it attractive and fascinating to scientific
minds.

The evidence in support of the nebulous origin of the solar system, if not conclusive, is of much weight and importance.   The remarkable harmony with which the orbs of the system perform their motions is strongly indicative of their common origin and that their evolution occurred in subordination to the law of universal gravitation.   The following are the characteristic points in favour of this theory :—

1. All the planets revolve round the Sun in the same direction, and they all occupy nearly the same plane.

2. Their satellites, with the exception of those of Uranus and Neptune, perform their revolutions in obedience to the same law.

3. The rotation on their axes of the Sun, planets, and satellites is in the same direction as their orbital motion.

Between the orbits of Mars and Jupiter there revolves a remarkable group of small planets or planetoids.   On account of the absence of a planet in this region, where, according to the laws of planetary distances, one ought to be found, the existence of those small bodies was suspected for some years prior to their discovery.   The first was detected by Piazzi at Palermo in 1801 ; two others were discovered by Olbers in 1802 and 1807, and one by Harding in 1804.   For some time it was believed that no more planetoids existed, but in 1845 a fifth was detected by Hencke, and from that year until now upwards of 300 of those small bodies have been discovered. Their magnitudes are of varied extent; the diameter of the largest is believed not to exceed 450 miles, and that of the smaller ones from twenty to thirty miles. It was surmised at one time, when only a few of those bodies were known, that they were the fragments of a planet which met with some terrible catastrophe; but since the discovery of so many other planetoids this theory cannot be maintained. According to the nebular hypothesis, these bodies are the consolidated portions of a nebulous ring which remained separate instead of having coalesced into one mass so as to form a planet. The uniform condensation of the ring would result in the formation of a multitude of small planets similar to what are found between the orbits of Mars and Jupiter. In Saturn's ring we have a remarkable instance of annular consolidation in which the form of the ring has been preserved. The ring is believed to consist of myriads of minute bodies, each of which travels in an orbit of its own as it pursues its path round the planet; the close approximation and exceeding minuteness of those moving objects create the appearance of a solid continuous ring.

Though, by means of the nebular hypothesis, it is impossible to explain all the phenomena associated with the motions of the orbs which enter into the structure of the solar system, yet this does not detract much from the merits of the theory, the fundamental principles of which are based upon the evolution of the solar system from a rotating nebula.

The retrograde motions of the satellites of Uranus
and Neptune, the velocity of the inner Martian
moon, and other abnormalities in the system, have
not as yet been explained, but doubtless there are
reasons by which those peculiarities can be accounted for if they were only known, *'felix qui
potuit cognoscere causas omnium rerum.'*

No attempt has been made to supplant the
nebular hypothesis by any other theory of cosmical
evolution. Modern investigations and discoveries
have strengthened its position, and at present it is
the only means by which we can account for the
existence of the visible material universe by which
we are surrounded.

In the days when Milton lived—three hundred
years ago—the nocturnal heavens presented the
same appearance to an observer as they do at the
present time. The stars pursued their identical
paths, and looked down upon the Earth with the
same aspect of serene tranquillity, regardless of the
vicissitudes which affect the inhabitants of this
terrestrial sphere. The constellations that adorn
the celestial vault duly appeared in their seasons,

> and in the ascending scale
> Of Heaven the stars that usher evening rose.—iv. 354-55.

The winter glories of Orion, the scintillating
brilliancy of Sirius, and the spangled firmament,
bearing no impress of change or variation which
would lead one to conclude that the heavens were
other than eternal, attracted then, as now, the admiration of beholders.

Apart from the orbs which constitute the solar system, little was known of the sidereal heavens beyond the visual effect created by the nocturnal aspect of the star-lit sky. Though ancient philosophers hazarded an opinion that the stars were suns, they received but scant attention from early astronomers, by whom they were merely regarded as convenient fixed points which enabled them to determine with greater accuracy the positions of the planets and the paths traced out by them in the heavens. The Ptolemaists, who believed in the diurnal revolution of the spheres, assigned to the stars a very subordinate place in their cosmology, which was the one adopted by Milton ; and although Copernicus relegated them to their proper location in space, yet he had no clear conception of a universe of stars. Tycho Brahé, who declined to accept the Copernican theory, disbelieved that the stars were suns, and Galileo, who discovered the stellar nature of the Milky Way, remarked that the stars were not illumined by the Sun's rays in the same manner that the planets are, but expressed no opinion with regard to their physical constitution. It is only within the past fifty years that proof has been obtained of the real nature of the stars. By the spectroscopic analysis of their light it has been ascertained that the elements of matter which enter into their composition exist in a condition similar to what is found in the Sun. The stars are therefore suns, many of them surpassing in magnitude and brilliancy the great luminary of our system.

Though Milton makes frequent allusion to the magnificence of the starry heavens, we have no evidence that he regarded the stars as suns, nor does he refer to them as such in any part of his poem.[1] What impressed him most was their number and brilliancy, to which reference is made in the following passages :

> About him all the Sanctities of Heaven
> Stood thick as stars.—iii. 60–61.

> And sowed with stars the Heavens thick as a field.
> —vii. 358.

> Amongst innumerable stars, that shone
> Stars distant, but nigh hand seemed other worlds.
> —iii. 564–65.

> her reign
> With thousand lesser lights dividual holds,
> With thousand thousand stars, that then appeared
> Spangling the hemisphere.—vii. 381–84.

Milton describes the number of the fallen angels as

> an host
> Innumerable as the stars of night.—v. 744–45,

and the attention of Satan is directed by the archangel Uriel to the multitude of stars formed from the chaotic elements of matter :

> Numberless as thou seest, and how they move ;
> Each had his place appointed, each his course ;
> The rest in circuit walls this universe.—iii. 719–21.

---

[1] An expression in Book VIII. 148-49 would seem to indicate that this was inaccurate, but the lines

> ' and other suns perhaps
> With their attendant moons, thou wilt descry,'

are an allusion to the planets Jupiter and Saturn, whose satellites had been recently discovered.

Though Milton was doubtless familiar with the leading orbs of the firmament and knew their names, and the constellations in which they are situated, yet he makes no direct allusion to any of them in his poem. Neither Arcturus, which is mentioned in the Book of Job, nor Sirius, which attracted the attention of Homer, who compared the brightness of Achilles' armour to the dazzling brilliancy of the dog-star, finds a place in ' Paradise Lost.' And yet the superior magnitude and brilliancy of some stars when compared with those of others did not escape Milton's observation when, in describing the lofty eminence of Satan in heaven, prior to his fall, he represents him as

brighter once amidst the host
Of angels than that star the stars among.—vii. 132–33.

There is but one star to which Milton makes individual allusion, and, though not of any conspicuous brilliancy, yet it is one of much importance to astronomers—

the fleecy star that bears
Andromeda far off Atlantic seas
Beyond the horizon.—iii. 558–60.

This is α Arietis, the first point in the constellation of that name, which signifies the Ram, and from which the right ascensions of the stars are measured on the celestial sphere. In the time of Hipparchus the ecliptic intersected the celestial equator in Aries, which indicated the commencement of the astronomical year and the occurrence of the vernal equinox ; but, owing to precession, this

point is now 30° westward of Aries and in the constellation Pisces. The star was called Hamal by
the Arabs, signifying a sheep, and the animal
is represented as looking backwards.   Manilius
writes :—

> First Aries, glorious in his golden wool,
> Looks back and wonders at the mighty Bull.

Aries is associated with the legend of the Golden
Fleece, in quest of which Jason and his valiant crew
sailed in the ship 'Argo.' In the autumn, Andromeda
is situated above Aries, and would seem to be borne
by the latter, which accounts for Milton's description of the relative positions of those two constellations.

Milton alludes to the starry sphere in several
passages in his poem, and also mentions the starry
pole above which he soared in imagination up to
the Empyrean or Heaven of Heavens. His contemplation of the Galaxy must have impressed his mind
with the magnitude and extent of the sidereal universe, for he was aware that this luminous zone which
encircles the heavens consists of myriads of stars, so
remote as to be incapable of definition by unaided
vision.   Milton's description of this vast assemblage
of stars is worthy of its magnificence, and the purpose with which he poetically associates this glorified highway testifies to the sublimity of his thoughts
and to the originality of his genius.   In those parts
of his poem in which he describes the glories of the
celestial regions, and instances the beautiful phenomena associated with the individual orbs of the

firmament, we are able to perceive with what exquisite delight he beheld them all.

The invention of the telescope, and the important discoveries made by Kepler, Galileo, and Newton in the seventeenth century, were the means of effecting a rapid advance in the science of astronomy; but that branch of it known as sidereal astronomy was not then in existence. The star depths, owing to inadequate telescopic power, remained unexplored, and the secrets associated with those distant regions were inviolable, and lay beyond the reach of human knowledge. The physical constitution of the stars was unknown, nor was it ascertained with any degree of certainty that they were suns. The knowledge possessed by astronomers in those days was but meagre compared with what is now known of the sidereal heavens. Milton's astronomical knowledge, we find, was commensurate with what was known of the stellar universe, and this he has conspicuously displayed in his poem.

CHAPTER VIII

DESCRIPTION OF CELESTIAL OBJECTS MENTIONED IN 'PARADISE LOST'

THE SUN

THE surpassing splendour of the Sun, as compared with that of any of the other orbs of the firmament, is not more impressive than his stupendous magnitude, and the important functions which it is his prerogative to fulfil.   Situated at the centre of our system—of which he may be regarded as 'both eye and soul'—the orb has a diameter approaching 1,000,000 miles, and a mass 750 times greater than that of all the planets combined.   These, by his attractive power, he retains in their several paths and orbits, and even far distant Neptune acknowledges his potent sway.   With prodigal liberality he dispenses his vast stores of light and heat, which illumine and vivify the worlds circling around him, and upon the constant supply of which all animated beings depend for their existence.   Deprived of the light of the Sun, this world would be enveloped in perpetual darkness, and we should all miserably perish.

The Sun is distant from the Earth about 93,000,000 miles.   His diameter is 867,000 miles, or nearly four times the extent of the radius of the Moon's orbit. The mass of the orb exceeds that of the Earth 330,000 times, and in volume 1,305,000 times. The Sun is a sphere, and rotates on his axis from west to east in 25 days 8 hours. The velocity of a point at the solar equator is 4,407 miles an hour. The density of the Sun is only one-fourth that of the Earth, or, in other words, bulk for bulk, the Earth is four times heavier than the Sun. The force of gravity at the Sun's surface is twenty-seven times greater than it is on the Earth ; it would therefore be impossible for beings constituted as we are to exist on the solar surface.

The dazzling luminous envelope which indicates to the naked eye the boundary of the solar disc is called the PHOTOSPHERE. It is most brilliant at the centre of the Sun, and diminishes in brightness towards the circumference, where its luminosity is but one-fourth that of the central portion of the disc. The photosphere consists of gaseous vapours or clouds, of irregular form and size, separated by less brilliant interstices, and glowing white with the heat derived from the interior of the Sun. In the telescope the photosphere is not of uniform brilliancy, but presents a mottled or granular appearance, an effect created by the intermixture of spaces of unequal brightness. Small nodules of intense brilliance, resembling ' rice-grains,' but which, according to Nasmyth, are of a willow-leaf shape with pointed extremities, which form a network over portions of the photosphere, are sprinkled profusely over a more faintly luminous background.   These 'grains' consist of irregular rounded masses, having an area of several hundred miles.   By the application of a high magnifying power they can be resolved into 'granules'—minute luminous dots which constitute one-fifth of the Sun's surface and emit three-fourths of the light.   This granulation is not uniform over the surface of the photosphere ; in some parts it is indistinct, and appears to be replaced by interlacing filamentous bands, which are most apparent in the penumbræ of the spots and around the spots themselves.   The 'granules' are the tops of ascending masses of intensely luminous vapour ; the comparatively dark 'pores' consist of similar descending masses, which, having radiated their energy, are returning to be again heated underneath the surface of the photosphere.

In certain regions of the photosphere several dark patches are usually visible, which are called 'sun-spots.'   At occasional times they are almost entirely absent from the solar disc.   It has been observed that they occupy a zone extending from 10° to 35° north and south of the solar equator, but are not found in the equatorial and polar regions of the Sun.   A sun-spot is usually described as consisting of an irregular dark central portion, called the *umbra* ; surrounding it is an edging or fringe less dark, consisting of filaments radiating inwards called the *penumbra*.   Within the umbra there is sometimes seen a still darker spot, called the *nucleus*. The umbra is generally uniformly dark, but at times filmy luminous clouds have been observed floating over it. The nucleus is believed to be the orifice of a tubular depression in the floor of the umbra, prolonged downwards to an unknown depth. The penumbra is brightest at its inner edge, where the filaments present a marked contrast when compared

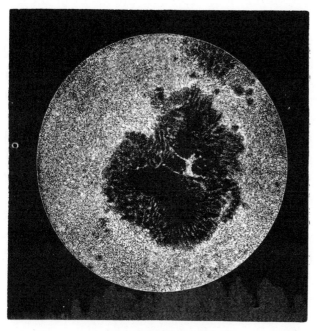

FIG. 6.—A Sun-spot magnified.
(*Janssen.*)

with the dark cavity of the umbra which they surround and overhang. Sometimes lengthened processes unite with those of the opposite side and form bands and ' bridges ' across the umbra. The darkest portion of the penumbra is its external edge, which stands out conspicuously against the adjoining bright surface of the Sun. One penumbra will sometimes enclose several umbræ whilst the nuclei may be entirely wanting.

Sun-spots usually appear in groups; large isolated spots are of rare occurrence, and are generally accompanied by several smaller ones of less perfect formation. The exact moment of the origin of a sun-spot cannot be ascertained, because it arises from an imperceptible point; it grows very rapidly, and often attains its full size in a day.

Prior to its appearance there is an unusual disturbance of the solar surface over the site of the spot: luminous ridges, called *faculæ*, and dark ' pores ' become conspicuous, between which greyish patches appear, that seem to lie underneath a thin layer of the photosphere; this is rapidly dispelled and a fully formed spot comes into view. When a sun-spot has completed its period of existence, the photospheric matter overwhelms the penumbra, and rushes into the umbra, which it obliterates, causing the spot to disappear. The duration of sun-spots is subject to considerable variation; some last for weeks or months, and others for a few days or hours. A spot when once fully formed maintains its shape, which is usually rounded, until the period of its breaking up. Spots of long duration rotate with the Sun. Those which become visible at the edge of the Sun's limb have been observed to travel across his disc in less than a fortnight, disappearing at the margin of the opposite limb; afterwards, if sufficiently long-lived, they have reappeared in twelve or thirteen days on the surface of the orb where first observed. It was by observation of the spots that the period of the axial rotation of the Sun became known.

Sun-spots vary very much in size—some are only a few hundred miles in width, whilst others have a diameter of 40,000 or 50,000 miles or upwards. In some instances the umbra alone has a breadth of 20,000 or 30,000 miles—three times the extent of the diameter of the Earth. Spots of this size are visible to the naked eye when the Sun is partially obscured by fog, or when his brilliancy is diminished by vapours near the horizon. A year seldom passes without the occurrence of several of such spots being recorded. The largest sun-spot ever observed had a diameter of about 150,000 miles. A group of spots, including their penumbræ, will occupy an area of many millions of square miles.

By long observation it has been ascertained that sun-spots increase and diminish in number with periodical regularity, and that a maximum sun-spot period occurs at the end of each eleven years. When spots are numerous on the Sun's disc there is great disturbance of the solar surface, accompanied by fierce rushes of intensely heated gases. This solar activity is known to influence terrestrial magnetism by causing a marked oscillation of the magnetic needle, and giving rise to so-called ' magnetic storms,' accompanied by magnificent displays of aurorae, with variations in electrical earth-currents. It would therefore appear that sun-spots have a

pronounced effect upon magnetic terrestrial phenomena, but how this is produced remains unknown.

Besides sun-spots, there are seen on the solar
disc bright flocculent streaks or ridges of luminous
matter called *faculæ*; they are found over the
whole surface of the Sun, but are most numerous
near the limb and in the immediate vicinity of
the spots. They have been compared to immense
waves—vast upheavals of photospheric matter, indicative of enormous pressure, and often extending in
length for many thousands of miles.

Nearly all observers have arrived at the conclusion that sun-spots are depressions or cavities in the
photosphere, but considerable difference of opinion
exists as to how they are formed. The most commonly
accepted theory is that they are caused by the pressure
of descending masses of vapour having a reduced
temperature, which absorb the light and prevent it
reaching us. Our knowledge of the Sun is insufficient to admit of any accurate conclusion on this
point; though we are able to perceive that the surface of the orb is in a state of violent agitation and
perpetual change, yet his great distance and intense
luminosity prevent our capability of perceiving the
ultimate minuter details which go to form the
texture of the solar surface. ' Bearing in mind that
a second of arc on the Sun represents 455 miles, it
follows that an object 150 miles in diameter is about
the *minimum visible* even as a mere mathematical
point, and that anything that is sufficiently large to
give the slightest impression of shape and extension

of surface must have an area of at least a quarter of a million square miles ; ordinarily speaking, we shall not gather much information about any object that covers less than a million.' [1] Since the British Islands have only an area of 120,700 square miles, it is evident that on the surface of the Sun there are many phenomena and physical changes occurring which escape our observation. Though the changes which occur in the spots and faculæ appear to be slow when observed through the telescope, yet in reality they are not so. Tremendous storms and cyclones of intensely heated gases, which may be compared to the flames arising from a great furnace, sweep over different areas of the Sun with a velocity of hundreds of miles an hour. Vast ridges and crests of incandescent vapour are upheaved by the action of internal heat, which exceeds in intensity the temperature at which the most refractory of terrestrial substances can be volatilised ; and downrushes of the same photospheric matter take place after it has parted with some of its stores of thermal energy. Sun-spots of considerable magnitude have been observed to grow rapidly and then disappear in a very short period of time ; occasionally a spot is seen to divide into two or more portions, the fragments flying asunder with a velocity of not less than 1,000 miles an hour. It is by these upheavals and convulsions of the solar atmosphere that the light and heat are maintained which illumine and vivify the worlds that gravitate round the Sun.

[1] Mr. E. W. Maunder, in *Knowledge*, March 1894.

During total eclipses of the Sun, several phenomena become visible which have enabled
astronomers to gain some further knowledge of the
nature of the solar appendages. The most important of these is the CHROMOSPHERE, which consists
of layers of incandescent gases that envelop the
photosphere and completely surround the Sun. Its
average depth is from 5,000 to 6,000 miles, and
when seen during an eclipse is of a beautiful rose
colour, resembling a sheet of flame. As seen in
profile at the edge of the Sun's disc, it presents an
irregular serrated appearance, an effect created by
the protuberance of luminous ridges and processes
—masses of flame which arise from over its entire
surface. The chromosphere consists chiefly of
glowing hydrogen, and an element called *helium*,
which has been recently discovered in a terrestrial
substance called cleveite ; there are also present the
vapours of iron, calcium, cerium, titanium, barium,
and magnesium. From the surface of this ocean of
fire, jets and pointed spires of flaming hydrogen
shoot up with amazing velocity, and attain an altitude of ten, twenty, fifty, and even one hundred
thousand miles in a very short period of time. They
are, however, of an evanescent nature, change rapidly
in form and appearance, and often in the course of an
hour or two die down so as not to be recognisable.
These *prominences*, as they are called, have been
divided into two classes. Some are in masses that
float like clouds in the atmosphere, which they
resemble in form and appearance ; they are usually

attached to the chromosphere by a single stem, or by slender columns ; occasionally they are entirely free. These are called *quiescent* prominences ; they consist of clouds of hydrogen, and are of more lasting duration than the other variety, called *eruptive* or metallic prominences. The latter are usually found in the vicinity of sun-spots, and, besides hydrogen, contain the vapours of various metals. They are of different forms, and present the appearance of filaments, spikes, and jets of liquid fire ; others are pyramidal, convoluted, and parabolic.

These outbursts, bending over like the jets from a fountain, and descending in graceful curves of flame, ascend from the surface of the chromosphere with a velocity often exceeding 100 miles in a second, and frequently reach an enormous height, but are of transient duration. They are closely connected with sun-spots, and are evidence of the tremendous forces that are in action on the surface of the Sun.

The CORONA is an aureole of light which is seen to surround the Sun during a total eclipse. It is an impressive and beautiful phenomenon, and is only visible when the Sun is concealed behind the dark body of the Moon. Professor Young gives the following graphic description of the corona : ' From behind it [the Moon] stream out on all sides radiant filaments, beams, and sheets of pearly light, which reach to a distance sometimes of several degrees from the solar surface, forming an irregular stellate halo, with the black globe of the

Moon in its apparent centre. The portion nearest
the Sun is of dazzling brightness, but still less
brilliant than the prominences, which blaze through
it like carbuncles. Generally this inner corona

Fig. 7.—The Corona during the Eclipse of May 1883.

has a pretty uniform height, forming a ring three
or four minutes of arc in width, separated by a
somewhat definite outline from the outer corona,
which reaches to a much greater distance and is

far more irregular in form. Usually there are several " rifts," as they have been called, like narrow beams of darkness, extending from the very edge of the Sun to the outer night, and much resembling the cloud shadows which radiate from the Sun before a thundershower. But the edges of these rifts are frequently curved, showing them to be something else than real shadows. Sometimes there are narrow bright streamers as long as the rifts, or longer. These are often inclined, or occasionally even nearly tangential to the solar surface, and frequently are curved. On the whole, the corona is usually less extensive and brilliant over the solar poles, and there is a recognisable tendency to accumulation above the middle latitudes, or spot zones ; so that, speaking roughly, the corona shows a disposition to assume the form of a quadrilateral or four-rayed star, though in almost every individual case this form is greatly modified by abnormal streamers at some point or other.' The corona surrounds the Sun and its other envelopes to a depth of many thousands of miles. It consists of various elements which exist in a condition of extreme tenuity; hydrogen, helium, and a substance called coronium appear to predominate, whilst finely divided shining particles of matter and electrical discharges resembling those of an aurora assist in its illumination.

We possess no knowledge of the physical structure of the interior of the Sun, nor have we any terrestrial analogy to guide us as to how matter

would behave when subjected to such conditions of
extreme temperature and pressure as exist in the
interior of the orb.   Yet we are justified in con-
cluding that the Sun is mainly a gaseous sphere
which is slowly contracting, and that the energy
expended in this process is being transformed into
heat so extreme as to render the orb a great fountain of light.

Milton in his poem makes more frequent allusion
to the Sun than to any of the other orbs of the
firmament, and, in all his references to the great
luminary, describes him in a manner worthy of his
unrivalled splendour, and of his supreme importance
in the system which he upholds and governs.
After having alighted on Mount Niphates, Satan is
described as looking

> Sometimes towards Heaven and the full-blazing Sun,
> Which now sat high in his meridian tower.—iv. 29–30.

He then addresses him thus :—

> O thou that with surpassing glory crowned,
> Look'st from thy sole dominion like the god
> Of this new World—at whose sight all the stars
> Hide their diminished heads—to thee I call,
> But with no friendly voice, and add thy name,
> O Sun, to tell thee how I hate thy beams,
> That bring to my remembrance from what state
> I fell, how glorious once above thy sphere.—iv. 32–39.

On another occasion :—

> The golden Sun in splendour likest Heaven
> Allured his eye.—iii. 572–73.

In describing the different periods of the day,
Milton seldom fails to associate the Sun with these

times, and rightly so, since they are brought about by the apparent diurnal journey of the orb across the heavens. Commencing with morning, he says :—

> Meanwhile,
> To re-salute the world with sacred light,
> Leucothea waked, and with fresh dews embalmed
> The Earth.—xi. 133–36.

> Soon as they forth were come to open sight
> Of day-spring, and the Sun—who, scarce up-risen,
> With wheels yet hovering o'er the ocean-brim,
> Shot parallel to the Earth his dewy ray,
> Discovering in wide landskip all the east
> Of Paradise and Eden's happy plains.—v. 138–43

> or some renowned metropolis
> With glistering spires and pinnacles adorned,
> Which now the rising Sun gilds with his beams.
> iii. 549–51.

> while now the mounted Sun
> Shot down direct his fervid rays, to warm
> Earth's inmost womb.—v. 300–302.

> for scarce the Sun
> Hath finished half his journey, and scarce begins
> His other half in the great zone of Heaven.—v. 558–60.

> To sit and taste, till this meridian heat
> Be over, and the Sun more cool decline.—v. 369–70.

> And the great Light of Day yet wants to run
> Much of his race, though steep. Suspense in Heaven,
> Held by thy voice, thy potent voice he hears,
> And longer will delay, to hear thee tell
> His generation, and the rising birth
> Of Nature from the unapparent deep.—vii. 98–103.

The declining day and approach of evening are described as follows :—

> Meanwhile in utmost longitude, where Heaven
> With Earth and Ocean meets, the setting Sun
> Slowly descended, and with right aspect
> Against the eastern gate of Paradise
> Levelled his evening rays.—iv. 539–43.

>   the Sun now fallen
> Beneath the Azores ; whether the Prime Orb,
> Incredible how swift, had thither rolled
> Diurnal, or this less volubil Earth,
> By shorter flight to the east, had left him there
> Arraying with reflected purple and gold
> The clouds that on his western throne attend.
>   iv. 591–97.

>   the parting Sun
> Beyond the Earth's green Cape and verdant Isles
> Hesperian sets, my signal to depart.—viii. 630–32.

> Now was the Sun in western cadence low
> From noon, and gentle airs due at their hour
> To fan the Earth now waked, and usher in
> The evening cool.—x. 92–95.

>   for the Sun,
> Declined, was hasting now with prone career
> To the Ocean Isles, and in the ascending scale
> Of Heaven the stars that usher evening rose.
>   iv. 352–55.

In the combat between Michael and Satan,
which ended in the overthrow of the rebel angels,
Milton, in his description of their armour, says :—

>   two broad suns their shields
> Blazed opposite.—vi. 305–306,

and in describing the faded splendour of the ruined
Archangel, the poet compares him to the Sun
when seen under conditions which temporarily
deprive him of his dazzling brilliancy and glory :—

> as when the Sun new-risen
> Looks through the horizontal misty air
> Shorn of his beams, or, from behind the Moon
> In dim eclipse, disastrous twilight sheds
> On half the nations, and with fear of change
> Perplexes monarchs.—i. 594-99.

This passage affords us an example of the sublimity of Milton's imagination and of his skill in adapting the grandest phenomena in Nature to the illustration of his subject.

THE MOON

The Moon is the Earth's satellite, and next to the Sun is the most important of the celestial orbs so far as its relations with our globe are concerned. Besides affording us light by night, the Moon is the principal cause of the ebb and flow of the tide—a phenomenon of much importance to navigators. The Moon is almost a perfect sphere, and is 2,160 miles in diameter. The form of its orbit is that of an ellipse with the Earth in the lower focus. It revolves round its primary in 27 days 7 hours, at a mean distance of 237,000 miles, and with a velocity of 2,273 miles an hour. Its equatorial velocity of rotation is 10 miles an hour. The density of the Moon is 3·57 that of water, or 0·63 that of the Earth ; eighty globes, each of the weight of the Moon, would be required to counterbalance the weight of the Earth, and fifty globes of a similar size to equal it in dimensions. The orb rotates on its axis in the same period of time in

which it accomplishes a revolution of its orbit ;
consequently the same illumined surface of the
Moon is always directed towards the Earth.   To
the naked eye the Moon appears as large as the
Sun, and it very rapidly changes its form and position in the sky.   Its motions, which are of a very
complex character, have been for many ages the
subject of investigation by mathematicians and
astronomers, but their difficulties may now be
regarded as having been finally overcome.

The phases of the Moon are always interesting
and very beautiful.   The orb is first seen in the
west, after sunset, as a delicate slender crescent of
pale light ; each night it increases in size, whilst it
travels eastward, until it attains the figure of a half
moon ; still growing larger as it pursues its course,
it finally becomes a full resplendent globe, rising
about the time that the Sun sets and situated directly opposite to him.   Then, in a reverse manner,
after full moon, it goes through the same phases,
until, as a slender crescent, it becomes invisible in
the solar rays ; afterwards to re-appear in a few
days, and, in its monthly round, to undergo the
same cycle of changes.   The phases of the Moon
depend upon the changing position of the orb with
regard to the Sun.   The Moon shines by reflected
light derived from the Sun, and as one half of its
surface is always illumined and the other half
totally dark, the crescent increases or diminishes
when, by the Moon's change of position, we see
more or less of the bright side.   Visible at first as a

slender crescent near the setting Sun, the angular distance from the orb and the width of the crescent increase daily, until, at the expiration of seven days, the Moon is distant one quarter of the circumference of the heavens from the Sun. The Moon is then a semi-circle, or in quadrature. At the end of other seven days, the distance of the Moon from the Sun is at its greatest—half the circumference of its orbit. It is then visible as a circular disc and we behold the orb as full moon. The waning Moon, as it gradually decreases, presents the same aspects reversed, and, finally, its slender crescent disappears in the Sun's rays. The convex edge of the crescent is always turned towards the Sun. The rising of the Moon in the east and its setting in the west is an effect due to the diurnal rotation of the Earth on her axis, but the orb can be perceived to have two motions besides : one from west to east, which carries it round the heavens in 29·53 days, and another from north to south. The west to east motion is steady and continuous, but, owing to the Sun's attractive force, the Moon is made to swerve from its path, giving rise to irregularities of its motion called PERTURBATIONS. The most important of these is the *annual equation*, discovered by Tycho Brahé—a yearly effect produced by the Sun's disturbing influence as the Earth approaches or recedes from him in her orbit ; another irregularity, called the *evection*, is a change in the eccentricity of the lunar orbit, by which the mean longitude of the Moon is increased or diminished. *Elliptic inequality, parallactic inequality,* the *variation,* and *secular acceleration,* are other perturbations of the lunar motion, which depend directly or indirectly on the attractive influence of the Sun and the motion of the Earth in her orbit.

As the plane of the Moon's orbit is inclined at an angle of rather more than 5° to the ecliptic, it follows that the orb, in its journey round the Earth, intersects this great circle at two points called the ' Nodes.' When crossing the ecliptic from south to north the Moon is in its ascending node, and when crossing from north to south in its descending node. In December the Moon reaches the most northern point of its course, and in June the southernmost. Consequently we have during the winter nights the greatest amount of moonlight, and in summer the least. In the evenings the moonlight is least in March and greatest in September, when we have what is called the Harvest Moon.

The telescopic appearance of the Moon is very interesting and beautiful, especially if the orb is observed when waxing and waning. As no aqueous vapour or cloud obscures the lunar surface, all its details can be perceived with great clearness and distinctness. Indeed, the topography of the Moon is better known than that of the Earth, for the whole of its surface has been mapped and delineated with great accuracy and precision. The Moon is in no sense a duplicate of its primary, and no analogy exists between the Earth and her satellite.

Evidence is wanting of the existence of an atmosphere surrounding the Moon; no clouds or exhalations can be perceived, and no water is believed to exist on the lunar surface. Consequently there are no oceans, seas, rivers, or lakes; no fertile plains or forest-clad mountains, such as are found upon the Earth. Indeed, all the conditions essential for the support and maintenance of organic life by which we are surrounded appear to be non-existent on the Moon. Our satellite has no seasons; its axial rotation is so slow that one lunar day is equal in length to fourteen of our days; this period of sunshine is succeeded by a night of similar duration. The alternation of such lengthened days and nights subjects the lunar surface to great extremes of heat and cold.

When viewed with a telescope, the surface of the Moon is perceived to consist of lofty mountain chains with rugged peaks, numerous extinct volcanoes called crater mountains, hills, clefts, chasms, valleys, and level plains—a region of desolation, presenting to our gaze the shattered and upturned fragments of the Moon's crust, convulsed by forces of a volcanic nature which have long since expended their energies and died out. The mountain ranges on the Moon resemble those of the Earth, but they have a more rugged outline, and their peaks are more precipitous, some of them rising to a height of 20,000 feet. They are called the Lunar Alps, Apennines, and Cordilleras, and embrace every variety of hill, cliff, mound, and ridge of comparatively low elevation.

The plains are large level areas, which are situated on various parts of the lunar surface ; they are of a darker hue than the mountainous regions by which they are surrounded, and were at one time believed to be seas. They are analogous to the prairies, steppes, and deserts of the Earth.

Valleys.—Some of these are of spacious dimensions ; others are narrow, and contract into gorges and chasms. Clefts or rills are long cracks or fissures of considerable depth, which extend sometimes for hundreds of miles across the various strata of which the Moon's crust is composed.

The characteristic features of the Moon's surface are the crater mountains : they are very numerous on certain portions of the lunar disc, and give the Moon the freckled appearance which it presents in the telescope, and which Galileo likened to the eyes in the feathers of a peacock's tail. They are believed to be of volcanic origin, and have been classified as follows : ' Walled plains, mountain rings, ring plains, crater plains, craters, craterlets, and crater cones.' Upwards of 13,000 of these mountains have been enumerated, and 1,000 are known to have a diameter exceeding nine miles. Walled plains consist of circular areas which have a width varying from 150 miles to a few hundred yards. They are enclosed by rocky ramparts, whilst the centre is occupied by an elevated peak. The depth of these formations, which are often far below the level of the Moon's surface, ranges from 10,000 to 20,000 feet. Mountain rings, ring plains, and crater plains resemble those already described, but are on a smaller scale ; the floors of the larger ones are frequently occupied by craters and craterlets. The latter exist in large numbers, and some portions of the Moon's surface appear honeycombed with them, the smaller craters resting on the sides of larger ones and occupying the bottoms of the more extensive areas. There is no kind of formation on the Earth's surface that can be compared with these crater mountains, which indicate that the Moon was at one time a fiery globe convulsed by internal forces which found an outlet in the numerous volcanoes scattered over her surface.

The most remarkable of these volcanic mountains have been named after distinguished men. (1) Copernicus is one of the most imposing ; its crater is 56 miles in diameter, and situated at its centre is a mountain with six peaks 2,400 feet in height. The ring by which it is surrounded rises 11,000 feet above the floor of the crater, and consists of terraces believed to have been created by the partial congelation and periodic subsidence of a lake of molten lava which occupied the enclosed area.

(2) Tycho is one of the most magnificent and perfect of lunar volcanoes, and is also remarkable as being a centre from which, when the Moon is full, there radiates a number of bright streaks which extend across the lunar surface, over mountain and valley, through ring and crater, for many hundreds of miles. Their nature is unknown, and nothing resembling them is found on the Earth.

Tycho has a diameter of 50 miles and a depth of 17,000 feet. The peak which rises from the floor of the crater attains a height of 6,000 feet, and the rampart consists of a series of terraces which give variety to the appearance of the inner wall. The surface of the Moon round Tycho is honeycombed with small volcanoes.

(3) Clavius is one of the most extensive of the walled plains; it has a diameter of 142 miles and an area of 16,500 square miles. The rocky annulus which surrounds it is very lofty and precipitous, and at one point reaches a height of 17,300 feet. Upwards of 90 craters have been counted within this space, one of the peaks attaining to an elevation of 24,000 feet above the level floor of the plain. It is believed that the lowest depths of this wild and precipitous region are never penetrated by sunlight, they are so overshadowed by towering crag and fell which intercept the solar rays; and, as there is no atmosphere to cause reflection, they are consequently enveloped in perpetual darkness.

(4) Plato has a diameter of about 60 miles and an area of 2,700 square miles; its central peak rises to a height of 7,300 feet. It has an irregular rampart which is broken up into terraces averaging about 4,000 feet high; three cones, each with an elevation of from 7,000 to 9,000 feet, rest on its western border.

(5) Theophilus is the deepest of the visible craters on the Moon. It has a diameter of 64 miles, and the inner edge of the ring rises

from the level floor to a height ranging from 14,000 to 18,000 feet. A group of mountains occupies the centre of the area, the highest peak of which reaches an elevation of 5,200 feet. Cyrillus and Catharina, two adjacent craters, are each about 16,000 feet deep and connected by a wide valley.

(6) Aristarchus is the brightest spot on the Moon, and appears almost dazzling in the telescope. The crater has a diameter of 42 miles, the centre of which is occupied by a steep mountain. The rampart on the western side rises to a height of 7,500 feet, on the east it becomes a plateau which connects it with a smaller crater called Herodotus. Bright streaks radiate from Aristarchus when there is full moon, and extend for a considerable distance over the surface of the orb.

Though the face of the Moon has been carefully scanned for two centuries and a half, and seleno-graphers have mapped and delineated her features with the utmost accuracy and precision, yet no perceptible change of a reliable character has been perceived to occur on any part of the orb. The surface of the hemisphere directed towards the Earth appears to be an alternation of desert plains, craggy wildernesses, and extinct volcanoes—a region of desolation unoccupied by any living thing, and 'upon which the light of life has never dawned.' Owing to the absence of an atmosphere, there is neither diffuse daylight nor twilight on the Moon. Every portion of the lunar surface not exposed to the Sun's rays is shrouded in darkness, and black shadows can be observed fringing prominences of silvery whiteness. If the Moon were enveloped in an atmosphere similar to that which surrounds the Earth, the reflection and diffusion of light among the minute particles of watery vapour which permeate it would give rise to a gradual transition from light to darkness; the lunar surface would be visible when not illumined by the direct rays of the Sun, and before sunrise and after sunset, dawn and twilight would occur as upon the Earth. But upon the Moon there is no dawn, and the darkness of night envelops the orb until the appearance of the edge of the Sun's disc above the horizon, then his dazzling rays illumine the summits and loftiest peaks of the lunar mountains whilst yet their sides and bases are wrapped in deep gloom. Since the pace of the Sun across the lunar heavens is 28 times slower than it is with us, there is continuous sunshine on the Moon for 304 hours, and this long day—equal to about a fortnight of our time—is succeeded by a night of similar duration. As there is no atmosphere overhead to diffuse or reflect the light, the Sun shines in a pitch-black sky, and at lunar noonday the planets and constellations can be seen displaying a brilliancy of greater intensity than can be perceived on Earth during the darkest night. Every portion of the Moon's surface is bleak, bare, and untouched by any softening influences. No gentle gale ever sweeps down her valleys or disturbs the dead calm that hangs over this world; no cloud ever tempers the fierce

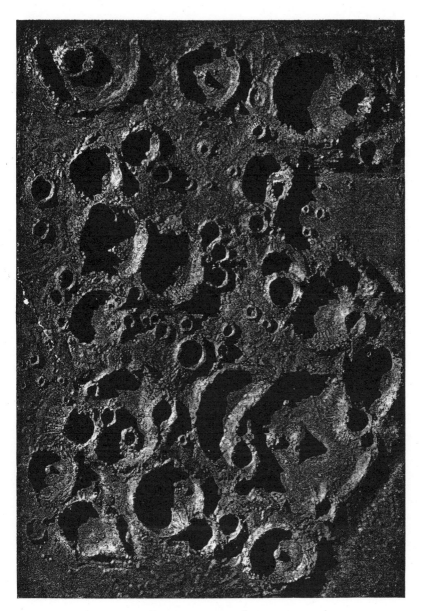

A PORTION OF THE MOON'S SURFACE

glare of the Sun that pours down his unmitigated rays from a sky of inky blackness; no refreshing shower ever falls upon her arid mountains and plains; no sound ever breaks the profound stillness that reigns over this realm of solitude and desolation.

As might be expected, Milton makes frequent allusion to the Moon in 'Paradise Lost,' and does not fail to set forth the distinctive charms associated with the unrivalled queen of the firmament. The majority of poets would most likely regard a description of evening as incomplete without an allusion to the Moon. Milton has adhered to this sentiment, as may be perceived in the following lines :—

> till the Moon,
> Rising in clouded majesty, at length
> Apparent queen, unveiled her peerless light,
> And o'er the dark her silver mantle threw.—iv. 606–609.

> now reigns
> Full-orbed the Moon, and with more pleasing light,
> Shadowy sets off the face of things.—v. 41–43.

The association of the Moon with the nocturnal revels and dances of elves and fairies is felicitously expressed in the following passage :—

> or faëry elves,
> Whose midnight revels, by a forest side
> Or fountain, some belated peasant sees,
> Or dreams he sees, while overhead the Moon
> Sits arbitress, and nearer to the Earth
> Wheels her pale course.—i. 781–86.

In contrast with this, we have Milton's description of the Moon when affected by the demoniacal practices of the 'night-hag' who was believed to destroy infants for the sake of drinking their blood, and applying their mangled limbs to the purposes of incantation. The legend is of Scandinavian origin and the locality Lapland :—

> Nor uglier follow the night-hag, when called
> In secret, riding through the air she comes,
> Lured with the smell of infant blood, to dance
> With Lapland witches, while the labouring Moon
> Eclipses at their charms.—ii. 662-66.

In his description of the massive shield carried by Satan, the poet compares it with the full moon :—

> his ponderous shield
> Ethereal temper, massy, large, and round,
> Behind him cast.   The broad circumference
> Hung on his shoulders like the Moon.—i. 284-87.

The phases displayed by the Moon in her monthly journey round the Earth, and which lend a variety of charm to the appearances presented by the orb, are poetically described by Milton in the following lines : —

> but there the neighbouring Moon
> (So call that opposite fair star) her aid
> Timely interposes, and her monthly round
> Still ending, still renewing, through mid-Heaven
> With borrowed light her countenance triform
> Hence fills and empties, to enlighten the Earth,
> And in her pale dominion checks the night.
>
> iii. 726-32

It is interesting to observe how aptly Milton describes the subdued illumination of the Moon's reflected light, as compared with the brilliant radiance of the blazing Sun, and how the distinguishing glory peculiar to each orb is appropriately set forth in the various passages in which they are described; their contrasted splendour enhancing rather than detracting from the grandeur and beauty belonging to each.

THE PLANET EARTH [1]

No lovelier planet circles round the Sun than the planet Earth, with her oceans and continents, her mountains, valleys, rivers, lakes, and plains; surrounded by heaven's azure, radiant with the sunlight of her day and adorned by night with countless sparkling points of gold. This beautiful world, the abode of MAN, is of paramount importance to us, and is the only part of the universe of which we have any direct knowledge.

The Earth may be regarded as one of the Sun's numerous family, and is situated third in order from the refulgent orb, round which it revolves in an elliptical orbit at a mean distance of 92,800,000 miles. The Earth is nearest to the Sun at the end of December, and furthest away at the beginning of July; the difference between those distances is 3,250,000 miles—the extent of the eccentricity of the planet's orbit. The figure of the Earth is that of an oblate spheroid; it is slightly flattened at the poles and bulges at the equator. Its polar or

[1] Though not a celestial body, it is considered desirable to describe the Earth as a member of the solar system.

shortest diameter is 7,899 miles, its equatorial
diameter is 7,926 miles—greater than the other by
27 miles.   The circumference of the Earth at the
equator is 24,899 miles, and the total area of its
surface is 197,000,000 square miles.   Its mean
density is 5½ times greater than that of water.

The two principal motions performed by the
Earth are : (1) Rotation on its axis ; (2) its annual
revolution round the Sun.   The Earth always
rotates in the same manner, and in the same direction, from west to east.   As the axis of rotation
corresponds with the shortest diameter of the
planet, it affords strong evidence that the Earth
assumed its present shape whilst rapidly rotating
round its axis when in a fluid or plastic condition.
This would accord with the nebular hypothesis.
The ends of the Earth's axis are called the poles
of the Earth ; one is the north, the other the south
pole.   The north pole is directed towards a star in
the Lesser Bear called the Pole Star.   The south
pole is directed to a corresponding opposite part of
the heavens.   The Earth's axis is inclined 63° 33′
to the plane of the ecliptic, and is always directed
to the same point in the heavens.   The Earth
accomplishes a revolution on its axis in 23 hours 56
minutes 4 seconds mean solar time, which is the
length of the sidereal day.   This rate of rotation is
invariable.   At the equator, where the circumference
of the globe exceeds 24,000 miles, the velocity of a
point on its surface is upwards of 1,000 miles an
hour, but, as the poles are approached, the tangential

velocity diminishes, and at those points it is entirely absent. The Earth accomplishes a revolution of her orbit in 365 days 6 hours 9 minutes; in her journey round the Sun she travels a circuit of 580,000,000 miles at an average pace of 66,000 miles an hour. The Earth has other slight motions called *perturbations*, which are produced by the gravitational attraction of other members of the solar system. The most important of these is Precession of the Equinoxes, which is caused by the attraction of the Sun, Moon, and planets, on the protuberant equatorial region of the globe. This attraction has a tendency to turn the Earth's axis at right angles to her orbit, but it only results in the slow rotation of the pole of the equator round that of the ecliptic, which is occurring at the rate of 1° in 70 years, and will require a period of 25,868 years to complete an entire revolution of the heavens.

The spot on Earth round which is centred the chief interest in Milton's poem is Paradise, which was situated in the east of Eden, a district of Central Asia. It was here where God ordained that man should first dwell—a place created for his enjoyment and delight. Satan, after his soliloquy on Mount Niphates, directs his way to Paradise, and arrives first in Eden, where he beholds from a distance the Happy Garden—

So on he fares, and to the border comes
Of Eden, where delicious Paradise,
Now nearer, crowns with her enclosure green,
As with a rural mound, the champain head
Of a steep wilderness, whose hairy sides

T

With thicket overgrown, grotesque and wild,
Access denied ; and overhead upgrew
Insuperable highth of loftiest shade,
Cedar, and pine, and fir, and branching palm,
A sylvan scene, and, as the ranks ascend,
Shade above shade, a woody theatre
Of stateliest view.   Yet higher than their tops
The verdurous wall of Paradise up-sprung ;
Which to our general sire gave prospect large
Into his nether empire neighbouring round.
And higher than that wall, a circling row
Of goodliest trees, loaden with fairest fruit,
Blossoms and fruits at once of golden hue,
Appeared, with gay enamelled colours mixed ;
On which the Sun more glad impressed his beams
Than in fair evening cloud, or humid bow,
When God hath showered the Earth : so lovely seemed
That landskip.   And of pure now purer air
Meets his approach, and to the heart inspires
Vernal delight and joy, able to drive
All sadness but despair.   Now gentle gales,
Fanning their odoriferous wings, dispense
Native perfumes, and whisper whence they stole
Those balmy spoils.—iv. 131–59.

Satan, having gained admission to the Garden
by overleaping the tangled thicket of shrubs and
bushes which formed an impenetrable barrier and
prevented any access to the enclosure within, he
flew up on to the Tree of Life—

Beneath him, with new wonder, now he views,
To all delight of human sense exposed,
In narrow room Nature's whole wealth ; yea, more !—
A Heaven on Earth : for blissful Paradise
Of God the garden was, by Him in the east
Of Eden planted, Eden stretched her line
From Auran eastward to the royal towers
Of great Seleucia, built by Grecian kings,

Or where the sons of Eden long before
Dwelt in Telassar. In this pleasant soil
His far more pleasant garden God ordained.
Out of the fertile ground he caused to grow
All trees of noblest kind for sight, smell, taste ;
And all amid them stood the Tree of Life,
High eminent, blooming ambrosial fruit
Of vegetable gold ; and next to life,
Our death, the Tree of Knowledge, grew fast by—
Knowledge of good, bought dear by knowing ill.
Southward through Eden went a river large,
Nor changed his course, but through the shaggy hill
Passed underneath ingulfed ; for God had thrown
That mountain, as his garden mould, high raised
Upon the rapid current, which, through veins
Of porous earth with kindly thirst up-drawn,
Rose a fresh fountain, and with many a rill
Watered the garden ; thence united fell
Down the steep glade, and met the nether flood,
Which from his darksome passage now appears,
And now, divided into four main streams,
Runs diverse, wandering many a famous realm
And country whereof here needs no account ;
But rather to tell how, if Art could tell
How, from that sapphire fount the crispèd brooks,
Rolling on orient-pearl and sands of gold,
With mazy error under pendent shades
Ran nectar, visiting each plant, and fed
Flowers worthy of Paradise, which not nice Art
In beds and curious knots, but Nature boon
Poured forth profuse on hill, and dale, and plain,
Both where the morning Sun first warmly smote
The open field, and where the unpierced shade
Imbrowned the noontide bowers.—iv. 205-46.

Milton's description of Paradise is not less remarkable in its way than the lurid scenes depicted
by him in Pandemonium. The versatility of his
poetic genius is nowhere more apparent than in the

charming pastoral verse contained in this part of
his poem. The poet has lavished the whole wealth
of his luxuriant imagination in his description of
Eden and blissful Paradise with its ' vernal airs '
and ' gentle gales,' its verdant meads, and murmur-
ing streams, ' rolling on orient-pearl and sands of
gold ; ' its stately trees laden with blossom and
fruit ; its spicy groves and shady bowers, over
which there breathed the eternal Spring.

In Book IX. Satan expresses himself in an
eloquent apostrophe to the primitive Earth, over
which he previously wandered for seven days—

> O Earth, how like to Heaven, if not preferred
> More justly, seat worthier of gods, as built
> With second thoughts, reforming what was old !
> For what God, after better, worse would build ?
> Terrestrial Heaven, danced round by other Heavens,
> That shine, yet bear their bright officious lamps,
> Light above light, for thee alone, as seems,
> In thee concentring all their precious beams
> Of sacred influence ! As God in Heaven
> Is centre, yet extends to all, so thou
> Centring receiv'st from all those orbs ; in thee,
> Not in themselves, all their known virtue appears,
> Productive in herb, plant, and nobler birth
> Of creatures animate with gradual life
> Of growth, sense, reason, all summed up in Man,
> With what delight I could have walked thee round,
> If I could joy in aught—sweet interchange
> Of hill and valley, rivers, woods, and plains,
> Now land, now sea, and shores with forest crowned,
> Rocks, dens, and caves.—ix. 99–118.

Though it is impossible to regard the Earth as
possessing the importance ascribed to it by the
ancient Ptolemaists ; nevertheless, our globe is a

great and mighty world, and appears to be one of the most favourably situated of all the planets, being neither near the Sun nor yet very far distant from the orb; and although, when compared with the universe, it is no more than a leaf on a tree in the midst of a vast forest; still, it is not the least important among other circling worlds, and unfailingly fulfils the part allotted to it in the great scheme of creation.

THE PLANET HESPERUS

This is the beautiful morning and evening star, the peerless planet that ushers in the twilight and the dawn, the harbinger of day and unrivalled queen of the evening. Venus, called after the Roman goddess of Love, and also identified with the Greek Aphrodite of ideal beauty, is the name by which the planet is popularly known; but Milton does not so designate it, and the name 'Venus' is not found in 'Paradise Lost.' The ancients called it Lucifer and Phosphor when it shone as a morning star before sunrise, and Hesperus and Vesper when it became visible after sunset. It is the most lustrous of all the planets, and at times its brilliancy is so marked as to throw a distinct shadow at night.

Venus is the second planet in order from the Sun. Its orbit lies between that of Mercury and the Earth, and in form approaches nearer to a circle than that of any of the other planets.

It travels round the Sun in 224·7 days, at a mean distance of 67,000,000 miles, and with an average velocity of 80,000 miles an hour. Its period of rotation is unknown. By the observation of dusky spots on its surface, it has been surmised that the planet completes a revolution on its axis in 23¼ hours; but other observers doubt this and are inclined to believe that it always presents the same face to the Sun. When at inferior conjunction Venus approaches nearer to the Earth than any other planet, its distance then being 27,000,000 miles. Its greatest elongation varies from 45° to 47° 12'; it therefore can never be much more than three hours above the horizon before sunrise, or after sunset. Venus is a morning star when passing from inferior to superior conjunction, and during the other half of its synodical period it is an evening star. The planet attains its greatest brilliancy at an elongation 40° west or east of the Sun—five weeks before and after inferior conjunction. It is at these periods, when at its greatest brilliancy, that it casts a shadow at night.

Though so pleasing an object to the unaided eye, Venus, when observed with the telescope, is often a source of disappointment—this is on account of its dazzling brilliancy, which renders any accurate definition of its surface impossible. Sir John Herschel writes: 'The intense lustre of its illuminated part dazzles the sight, and exaggerates every imperfection of the telescope; yet we see clearly that its surface is not mottled over with permanent spots like the Moon ; we notice in it neither mountains nor shadows, but a uniform brightness, in which sometimes we may indeed fancy, or perhaps more than fancy, brighter or obscurer portions, but can seldom or never rest fully satisfied of the fact.' It is believed that the surface of the planet is invisible on account of the existence of a cloud-laden atmosphere by which it is enveloped, and which may serve as a protection against the intense glare of the sunshine and heat poured down by the not far-distant Sun. Schröter, a German astronomer, believed that he saw lofty mountains on the surface of the planet, but their existence has not been confirmed by any other observer. The Sun if viewed from Venus would have a diameter nearly half as large again as when seen from the Earth ; it is therefore probable that the planet is subjected to a much higher temperature than what is experienced on our globe.

The phases of Venus are similar to those exhibited by the Moon, and are caused by a change in position of the illumined hemisphere of the planet with regard to the Earth. At superior conjunction the whole enlightened disc of the planet is turned towards the Earth, but is invisible by being lost in the Sun's rays. Shortly before or after it arrives at this point, its form is gibbous, the illumined portion being less than a circle but greater than a semicircle. At its greatest elongation west or east of the Sun the planet resembles the Moon in quadrature—a half moon—and between those points and inferior conjunction it is visible as a beautiful crescent. It becomes narrower and sharper as it approaches inferior conjunction, until it resembles a curved luminous thread prior to its disappearance at the conjunction. After having passed this point it reappears on the other side of the Sun as the morning star.

It would be only natural to imagine that this peerless orb, the most beautiful and lustrous of the planets, upon which men have gazed with long-ing admiration, and designated the emblem of ' all beauty and all love,' should have impressed Milton's poetical imagination with its charming appearance, and stimulated the flow of his capti-vating muse. He addresses the orb as

> Fairest of Stars, last in the train of night,
> If better thou belong not to the dawn,
> Sure pledge of day, that crown'st the smiling morn
> With thy bright circlet, praise Him in thy sphere
> While day arises, that sweet hour of prime.—v. 166–70

In these lines the poet alludes to Venus as the morning star.

In the other passages in his poem Milton asso-ciates the planet sometimes with the morning and at other times with the evening—

> His countenance, as the Morning Star that guides
> The starry flock.—v. 708–709.

> Or if the Star of Evening and the Moon
> Haste to thy audience, Night with her will bring
> Silence, and Sleep listening to thee will watch.
> vii. 104–106.

And hence the morning planet gilds her horns.—vii. 366.

The Sun was sunk and after him the Star
Of Hesperus, whose office is to bring
Twilight upon the Earth, short arbiter
Twixt day and night.—ix. 47–50.

and bid haste the Evening Star
On his hill top to light the bridal lamp.—viii. 519–20.

Milton knew of the phases of Venus and was aware that at certain times the planet was visible in the telescope as a beautiful crescent. The line in which he mentions her as gilding her horns is an allusion to this appearance of Venus.

THE PLEIADES

The beautiful cluster of the Pleiades or Seven Sisters has been regarded with hallowed veneration from time immemorial. The happy influences believed to be shed down upon the Earth by those stars and their close association with human destinies have rendered them objects of almost sacred interest among the different races of mankind. In every region of the globe and in every clime, among civilised nations and savage fetish-worshipping tribes, the same benign influences were ascribed to the stars which form this interesting group.

In Greek mythology they were known as the seven daughters of Atlas and Pleione. Different versions are given of their fate. By some writers it is said they died from grief in consequence of the death of their sisters, the Hyades, or on account of

the fate of their father, who, for treason, was condemned by Zeus to bear on his head and hands the
vault of heaven, on the mountains of north-west
Africa which bear his name.   According to others
they were the companions of Diana, and, in order
to escape from Orion, by whom they were pursued,
the gods translated them to the sky.

All writers agree in saying that after their death
or translation they were transformed into stars.
Their names are Alcyone, Electra, Maia, Merope,
Sterope, Taygeta, and Celaeno.   The seventh
Atlantid is said to be the 'lost Pleiad,' but it can
be perceived without difficulty by a person possessing good eyesight.   In the book of Job there
is a beautiful allusion to the Pleiades (chap.
xxxviii.) when God speaks out of the whirlwind
and asks the patriarch to answer Him—

Canst thou bind the sweet influences of the Pleiades, or loose
the bands of Orion ?
Canst thou bring forth Mazzaroth in his season ? or canst
thou guide Arcturus with his sons ?
Knowest thou the ordinances of heaven ? canst thou set
the dominion thereof in the earth ?

Admiral Smyth says that this noble passage is
more correctly rendered as follows :

Canst thou bind the delightful teemings of Cheemah ?
Or the contractions of Chesil canst thou open ?
Canst thou draw forth Mazzaroth in his season
Or Ayeesh and his sons canst thou guide ?

He writes : ' In this very early description of the
cardinal constellations, *Cheemah* denotes Taurus
with the Pleiades ; *Chesil* is Scorpio ; Mazzaroth is

Sirius in " the chambers of the south ; " and Ayeesh
the Greater Bear, the Hebrew word signifying
a *bier*, which was shaped by the four well-known
bright stars, while the three forming the tail were
considered as children attending a funeral.' The
Greeks at an early period were attracted by this
cluster of stars, and Hesiod alludes to them in his
writings. One passage converted into rhyme reads
as follows :

> There is a time when forty days they lie,
> And forty nights, conceal'd from human eye ;
> But in the course of the revolving year,
> When the swain sharps the scythe, again appear.

Their heliacal rising was considered a favourable
time for setting out on a voyage, and their midnight
culmination, which occurred shortly after the middle
of November, was celebrated by some nations with
festivals and public ceremonies. Considerable diversity of opinion existed among the ancients with
regard to the number of stars which constitute
this group. It was affirmed by some that only six
were visible, whilst others maintained that seven
could be seen. Ovid writes :

> Quae septem dici, sex tamen esse solent.

Homer and Attalus mention six ; Hipparchus and
Aratus seven. The legend with regard to the lost
Pleiad would seem to indicate that, during a period
in the past, the star possessed a superior brilliancy
and was more distinctly visible than it is at the
present time. This may have been so, for, should
it belong to the class of variable stars, there would

be a periodic ebb and flow of its light, by which its fluctuating brilliance could be explained. When looked at directly only six stars can be seen in the group, but should the eye be turned sideways more than this number become visible. Several observers have counted as many as ten or twelve, and it is stated by Kepler that his tutor, Maestlin, was able to enumerate fourteen stars and mapped eleven in their relative positions. With telescopic aid the number is largely increased—Galileo observed thirty-six with his instrument and Hooke, in 1664, counted seventy-eight. Large modern telescopes bring into view several thousand stars in this region.

The Pleiades are situated at a profound distance in space. Their light period is estimated at 250 years, indicating a distance of 1,500 billions of miles. Our Sun if thus far removed would be reduced to a tenth-magnitude star. 'There can be little doubt,' says Miss Agnes Clerke, 'that the solar brilliancy is surpassed by sixty to seventy of the Pleiades. And it must be in some cases enormously surpassed; by Alcyone 1,000, by Electra 480, by Maia nearly 400 times. Sirius itself takes a subordinate rank when compared with the five most brilliant members of a group, the real magnificence of which we can thus in some degree apprehend.' This is the only star cluster which can be perceived to be moving in space, or which has an ascertained common proper motion. Its constituents form a magnificent system in which the stars bear a mutual relationship to each other, and perform intricate internal revolutions, whilst they in systemic union drift along through the depths of space. There are two allusions to the Pleiades in ' Paradise Lost.' In describing the path of the newly created Sun, Milton introduces them as indicative of the joyfulness associated with the birth of the Universe—

> First in his east the glorious lamp was seen,
> Regent of day, and all the horizon round
> Invested with bright rays, jocund to run
> His longitude through heaven's high road ; the grey
> Dawn, and the Pleiades before him danced,
> Shedding sweet influence.—vii. 370–75.

It was believed that the Earth was created in the spring ; and towards the end of April this group rises a little before the Sun and precedes him in his course, ' shedding sweet influences.' The ancients believed that the good or evil influences of the stars were exercised not in the night but during the day, when their rays mingled with those of the Sun. The pernicious influence of the Dog-star is mentioned by Latin writers as being most pronounced during the dog-days, at the end of summer and commencement of autumn, the time of the heliacal rising of this star.

The other allusion to the Pleiades is in Book X., line 673, where Milton, in describing the altered path of the Sun consequent upon the Fall, mentions how the orb travels through Taurus with the Seven Atlantic Sisters—the seven daughters of Atlas, the Pleiades, which are situated on the shoulder of the animal representing this zodiacal constellation.

THE GALAXY

The Galaxy or Milky Way is the great luminous zone encircling the heavens, which can be seen extending across the sky from horizon to horizon. Its diffused nebulous appearance caused the ancients much perplexity, and many quaint opinions were hazarded as to the nature of this celestial highway ; but the mystery associated with it was not solved until Galileo directed his newly invented telescope to this lucent object, when, to his intense delight, he discovered that it consists of myriads of stars— millions upon millions of suns so distant as to be individually indistinguishable to ordinary vision, and so closely aggregated, that their blended light gives rise to the milky luminosity signified by its name. This stelliferous zone almost completely encircles the sphere, which it divides into two nearly equal parts, and is inclined at an angle of 63° to the celestial equator.   In Centaurus it divides into two portions, one indistinct and of interrupted continuity, the other bright and well defined ; these, after remaining apart for 120°, reunite in Cygnus.   The Milky Way is of irregular outline and varies in breadth from 5° to 16° ; it intersects the equinoctial in the constellations Monoceros and Aquila, and approaches in Cassiopeia to within 27° degrees of the north pole of the heavens ; an equal distance intervenes between it and the south pole.   Its poles are in Coma Bernices and Cetus.   The stars in the galactic tract are very unevenly distributed ;

in some of its richest regions as many stars as are visible to the naked eye on a clear night have been counted within the space of a square degree. In other parts they are much less numerous, and there have been observed besides, adjacent to the most luminous portions of the zone, dark intervals and winding channels almost entirely devoid of stars. An instance of this kind occurs in the constellation of the Southern Cross, where there exists in a rich stellar region a large oval-shaped dark vacuity, 8° by 5° in extent, that appears to be almost entirely denuded of stars. In looking at it, an impression is created that one is gazing into an empty void of space far beyond the Milky Way. This gulf of Cimmerian darkness was called by early navigators the Coal Sack. Similar dark spaces, though not of such magnitude, are seen in Ophiuchus, Scorpio, and Cygnus.

The Galaxy, when viewed with a powerful telescope, is found to consist of congeries of stars, vast stellar aggregations, great luminous tracts resolvable into clouds of stars of overpowering magnificence, superb clusters of various orders, and convoluted nebulous streams wandering ' with mazy error ' among ' islands of light and lakes of darkness,' resolved by the telescope into banks of shining worlds. The concourses of stars which enter into the formation of this wonderful zone exhibit in a marvellous degree the amazing profusion in which these orbs exist in certain regions of space; yet those multitudes of stars perform their motions in harmonious unison and in orderly array, and by their mutual attraction sustain the dynamical equilibrium of this stupendous galactic ring, the diameter of which, according to one authority, is not traversed by light in less than 13,000 years.

Sir William Herschel, to whom we are indebted for most of what we know of the Milky Way, commenced a series of observations in 1785 with the object of acquiring a knowledge of the structure of the sidereal heavens. In the accomplishment of this object, to which he devoted a considerable part of his life, he undertook a systematic survey of that portion of the Galaxy which is visible in the Northern Hemisphere. By a method called star-gauging, which consisted in the enumeration of the stars in each successive telescopic field as the instrument moved slowly over the region under observation, he found that the depth of the star strata could be approximately ascertained by counting the stars along the line of vision; those were most numerous where the visual line appeared of the greatest length and fewest in number where it was shortest. Herschel perceived the internal structure of the Galaxy to be exceedingly intricate and complex, and that it embraced within its confines an endless variety of systems, clusters, and groups, branches, sprays, arches, loops, and streaming filaments of stars, all of which combined to form this luminous zone. 'It is indeed,' says a well-known astronomer, 'only to the most careless glance, or when viewed through an atmosphere of imperfect transparency, that the Milky

Way seems a continuous zone.   Let the naked eye
rest thoughtfully on any part of it, and, if circumstances be favourable, it will stand out rather as an
accumulation of patches and streams of light of
every conceivable variety of form and brightness,
now side by side, now heaped on each other ; again

Fig. 8.—A Portion of the Milky Way.

spanning across dark spaces, intertwining and forming a most curious and complex network ; and at
other times darting off into the neighbouring skies
in branches of capricious length and shape which
gradually thin away and disappear.'   Sir John
Herschel, who was occupied for four years at the

U

Cape of Good Hope in exploring the celestial regions
of the Southern Hemisphere, describes the coming on
of the Milky Way as seen in his 20-foot reflector.
He first remarks 'that all the stars visible to us,
whether by unassisted vision or through the best
telescopes, belong to and form part of a vast stratum
or considerably flattened and unsymmetrical congeries of stars in which our system is deeply and
eccentrically plunged ; and, moreover, situated near
a point where the stratum bifurcates or spreads
itself out into two sheets.' 'As the main body of
the Milky Way comes on the frequency and variety
of those masses (nebulous) increases ; here the
Milky Way is composed of separate or slight or
strongly connected clouds of semi-nebulous light,
and, as the telescope moves, the appearance is that
of clouds passing in a scud, as sailors call it.'   The
Milky Way is like sand, not strewed evenly as with
a sieve, but as if flung down by handfuls (and both
hands at once), leaving dark intervals, and all consisting of stars of the fourteenth, sixteenth, twentieth magnitudes down to nebulosity, in a most as-
tonishing manner.   After an interval of comparative
poverty, the same phenomenon, and even more
remarkable, I cannot say it is nebulous, it is all
resolved, but the stars are inconceivably numerous
and minute ; there must be millions and all almost
equally massed together.   Yet they nowhere run to
nuclei or clusters much brighter in the middle.
Towards the end of the seventeenth hour (Right
Ascension) the globular clusters begin to come in ;

they consist of stars of excessive minuteness, but yet not more so than the ground of the Milky Way, on which not only they appear projected, but of which it is very probable they form a part. 'From the foregoing analysis of the telescopic aspect of the Milky Way in this interesting region, I think it can hardly be doubted that it consists of portions differing exceedingly in distance, but brought by the effect of projection into the same, or nearly the same, visual line; in particular, that at the anterior edge of what we have called the main stream, we see foreshortened a vast and illimitable area scattered over with discontinuous masses and aggregates of stars in the manner of the cumuli of a mackerel sky, rather than of a stratum of regular thickness and homogeneous formation.'

The profound distance at which the stars of the Galaxy are situated in space precludes the possibility of our obtaining any definite knowledge of their magnitude and of the extent of the intervals by which they are separated from each other, nor can we learn anything of the details associated with the systems and combinations into which they enter. It is believed that the majority of the stars in the Milky Way equal or surpass the Sun in brilliancy and splendour. They are tenth to fifteenth magnitude stars; now, the Sun at the distance indicated by these magnitudes would in the telescope appear a much fainter object; he would not reach the fifteenth magnitude. Consequently, the galactic stars are regarded as his peers or superiors in magnitude and brilliancy. Those myriads of suns are all in motion —in nature a stationary body is unknown—and they are sufficiently far apart so as not to be unduly influenced by their mutual gravitational attraction ; a distance perhaps equal to that which separates our Sun from the nearest fixed star may intervene between each of those orbs. In the deepest recesses of the Milky Way, Sir William Herschel was able to count 500 stars receding in regular order behind each other ; between each there existed an interval of space, probably not less extensive than the interstellar spaces among the stars by which we are surrounded.

The richest galactic regions in the Northern Hemisphere are found in Perseus, Cygnus, and Aquila. Night after night could be spent in sweeping the telescope over fields where the stars can be seen in amazing profusion. In the interval of a quarter of an hour, Sir William Herschel observed 116,000 stars pass before him in the telescope, and on another occasion he perceived 258,000 stars in the space of forty-one minutes. In the constellation of the Swan there is a region about 5° in breadth which contains 331,000 stars. Photography reveals in a remarkable manner the amazing richness of this stelliferous zone ; the impress of the stars on the sensitive plate of the camera, in some instances, resembles a shower of descending snowflakes.

Though Sir William Herschel was able to fathom the Galaxy in most of its tracts, yet there were regions which his great telescopes were unable to penetrate entirely through. In Cepheus there is a spot where he observed the stars become 'gradually less till they escape the eye so that appearances here favour the idea of a succeeding more distant clustering part.' He perceived another in Scorpio 'where, through the hollows and deep recesses of its complicated structure, we behold what has all the appearance of a wide and indefinitely prolonged area strewed over with discontinuous masses and clouds of stars which the telescope at length refuses to analyse.' The Great Cluster in Perseus, which lies in the Milky Way, also baffled the penetrative capacity of Herschel's instruments. We cannot help quoting Professor Nichol's description of Herschel's observation of this remarkable object. He says: 'In the Milky Way, thronged all over with splendours, there is one portion not unnoticed by the general observer, the spot in the sword-hand of Perseus. That spot shows no stars to the naked eye; the milky light which glorifies it comes from regions to which unaided we cannot pierce. But to a telescope of considerable power the space appears lighted up with unnumbered orbs; and these pass on through the depths of the infinite, until, even to that penetrating glass, they escape all scrutiny, withdrawing into regions unvisited by its power. Shall we adventure into these deeper retirements? Then, assume an instrument of higher efficacy, and lo! the change is only repeated; those scarce observed before appear as large orbs, and, behind, a new

series begins, shading gradually away, leading towards farther mysteries! The illustrious Herschel
penetrated on one occasion into this spot, until
he found himself among depths whose light could
not have reached him in much less than 4,000
years; no marvel that he withdrew from the
pursuit, conceiving that such abysses must be
endless!' The Milky Way may be regarded as a
universe by itself, and our Sun as one of its myriad
stars.

Milton was aware of the stellar constitution of
the Milky Way, which was one of Galileo's discoveries. The poet gives a singularly accurate description of this luminous path, which he glorifies
as the way by which the Deity returned up to the
Heaven of Heavens after He finished His great
work of creation—

                                    So sung
      The glorious train ascending : He through Heaven,
      That opened wide her blazing portals, led
      To God's eternal house direct the way—
      A broad and ample road, whose dust is gold,
      And pavement stars, as stars to thee appear
      Seen in the Galaxy, that Milky Way
      Which nightly as a circling zone thou seest
      Powdered with stars.—vii. 573–81.

COMETS

Records of the appearance of these remarkable
objects have been handed down from earliest times ;
and when one of those mysterious visitors, travelling from out the depths of space, became visible
in our skies, it was regarded with apprehension and

dread as betokening the occurrence of calamities and direful events among the nations of the Earth.

The word comet is derived from the Greek κόμη, signifying ' hair,' to which the hazy, luminous appearance of those objects bears some resemblance. A comet consists of a bright central part called the *nucleus*; this is surrounded by layers of nebulous matter called the *coma*, and both combined form the *head*, from which a long appendage extends called the *tail*. The nucleus and tail are not essential parts of a comet, for many have been observed in which both have been wanting. The tail is frequently very conspicuous, and presents considerable diversity both as regards its appearance and length. In some comets it is entirely absent, and in others it has been observed to stretch over an arc of sixty or seventy degrees, indicating a length of 100 to 150 million miles. Sometimes it is straight, and at other times it is curved at the extremity; it has been observed bifurcated into two branches; and, on rare occasions, comets have been seen with two or more tails. The tail of a comet is always directed away from the Sun; it increases in size as the comet approaches the orb, and diminishes as it recedes from him. This depends upon the degree of heat to which the comet is exposed, which has the effect of driving off or evaporating some of the matter composing the head. During the time the comet is travelling round the Sun there is a continuous emission of this highly attenuated matter, which is visible as

the tail, but when the comet begins to recede from
the orb and reaches cooler regions of space the tail
diminishes in size as the temperature becomes
reduced, and ultimately it disappears.

The appearance of a comet in the sky is often
sudden and unexpected, and one of those erratic
wanderers may become visible at any time and in
any part of the heavens. It was remarked by
Kepler that there are as many comets in the sky as
there are fishes in the ocean. This may or may
not be true, for they only become visible when they
approach the Sun, and the time during which they
remain so does not usually exceed a few weeks or
months. Ancient astronomers were much perplexed with the motions of comets, which appeared
to be much more irregular than those of other
celestial bodies and unconformed to any known
laws. Tycho Brahé believed that comets moved
in circular orbits, and Kepler imagined that they
travelled in straight lines outwards from the Sun.
Newton, however, was able to demonstrate that any
conic section can be described about the Sun consistent with the law of gravitation, and that the
orbits of comets correspond with three of the four
sections into which a cone can be divided. Consequently, they obey the laws of planetary motion.
Comets which move in ellipses of known eccentricity and return with periodical regularity may
be regarded as belonging to the solar system.
Twenty of these are known, and eleven of them
have more than once passed their perihelion.

Those most familiarly known complete their periods
in years as follows :—Encke's 3·3; Swift's, 5·5;
Winnecke's, 5·6; Tempel's, 6; Brorsen's, 5·5; Faye's,
7·4; Tuttle's, 13·8, and Halley's, 76. Comets with
parabolic and hyperbolic orbits may be regarded as
stray objects which visit our system once, and
depart never to return again. Besides those already mentioned there are many comets with
orbits of such marked eccentricity that their ellipses when near perihelion cannot be distinguished
from parabolæ. The great comets of 1780, 1811,
1843, 1858, 1861, and 1882 traverse orbits approaching this form, and some of them require
hundreds and thousands of years to accomplish a
circuit of their paths.

Numerous instances of the appearance of remarkable comets have been recorded in the annals
of ancient nations. The earliest records of comets
are by the Chinese, who were careful observers
of celestial phenomena. A comet is said to have
appeared at the time of the birth of Mithridates
(134 B.C.), which had a disc as large as that of the
Sun; a great comet also became visible in the
heavens about the time of the death of Julius
Cæsar (44 B.C.), and another was seen in the reign
of Justinian (531 A.D.). A remarkable comet was
observed in 1106, and in 1456, the year in which
the Turks obtained possession of Constantinople
and threatened to overrun Europe, a great comet
appeared, which was regarded by Christendom with
ominous forebodings. The celebrated astronomer

Halley was the first to predict the return of a comet. Having become acquainted with Newton's investigations, which showed that the forms of the orbits of comets were either parabolæ or extremely elongated ellipses, he subjected the next great comet, which appeared in 1682, to a series of observations, calculated its orbit, and predicted that it would return to perihelion in seventy-five or seventy-six years. On referring to past records he discovered that a great comet appeared in 1607, which pursued a path similar to the one traced out for his comet, another was seen in 1531, and one in 1456. Halley perceived that the intervals between those dates corresponded to a period of about seventy-six years, the time which he calculated would be required for his comet to complete a revolution of its orbit. He therefore had no hesitation in predicting that the comet would appear again in 1758. Halley knew that he would not be alive to witness the event, and alludes to it in the following sentence : ' Wherefore if it should return according to our prediction about the year 1758, impartial posterity will not refuse to acknowledge that this was first discovered by an Englishman.' As the time approached when the comet should be drawing near to our system, much interest was excited among astronomers, who would have an opportunity afforded them of testing the accuracy of Halley's prediction. An eminent French mathematician named Clairaut computed anew, by a method rather different to that adopted by Halley, the retarding effect of the attraction of the planets upon the speed of the comet, and arrived at the conclusion that it would reach perihelion about the middle of April 1759 ; but, owing to unknown influences—Uranus and Neptune not having been discovered—it might be a month before or behind the calculated time. Clairaut made this announcement on November 14, 1758. Astronomers were now intently on the look-out for the comet, and night after night the sky was swept by telescopes in search of the expected visitor, which for upwards of seventy years had been pursuing its solitary path invisible to mortal eyes. But the mental vision of the mathematician did not fail to follow this celestial object, which was now announced as being on the confines of our system. The comet was first observed on December 25, 1758, it soon became conspicuous in the heavens, and reached perihelion on March 12, 1759, a month before the time assigned to it by Clairaut but within the limit of error allowed for unknown influences. Halley's comet returned again in 1835, and may be expected about the year 1911. The periodic appearance of this comet has been traced back to the year 1305.

The celebrated comet of 1680 was noted as having been the one which afforded Newton an opportunity of making observations which led to his discovery that comets describe orbits round the Sun in conformity with the different sections of a cone. The comet of 1811 was observed for many

weeks in the northern heavens as a brilliant object
with a beautiful fan-shaped tail; it completes a
revolution of its orbit in about 3,000 years. The
comet of 1843 was also a splendid object. It possessed a tail 200 million miles in length, and
approached within 32,000 miles of the Sun. The
heat to which it was exposed was sufficient to vola-
tilize the most infusible substances known to exist.
Donati's comet of 1858 will be long remembered as
one of the most impressive of celestial spectacles:
its tail extended over an area of forty degrees, and
enveloped the star Arcturus, which could be seen
shining through it with undiminished brilliancy.
Its period is estimated to be 2,100 years. A great
comet appeared in 1861, through the tail of which
the Earth passed without any perceptible effect
having resulted. No remarkable comets have appeared during recent years. In 1880, 1881, and
1882, several were observed, and that of 1881 was
the first successfully photographed.

Comets consist of cosmical matter which exists
in a condition of extreme tenuity, and especially so
in the coma and tail. Sir John Herschel described
them as almost spiritual in texture, and small stars
have been seen shining through their densest parts
without any perceptible diminution of their light.
The nucleus is believed to be composed of a congeries of meteoric fragments, and these, when exposed
to the Sun's heat, throw off luminous nebulous particles that are swept by some repulsive force into
space and form the appendage known as the tail.

Comets may be regarded as celestial objects that are perfectly innocuous. Neither fear nor dread need be apprehended from their visits; they come to please and instruct, not to injure or destroy.

Milton does not fail to introduce into his poem several allusions to comets, and in doing so expresses the ideas and sentiments which in his time were associated with those objects.

In describing the hostile meeting between Satan and Death before the Gates of Hell, he writes:

> On the other side,
> Incensed with indignation, Satan stood
> Unterrified, and like a comet burned,
> That fires the length of Ophiuchus huge
> In the arctic sky, and from his horrid hair
> Shakes pestilence and war.—ii. 706–11.

This passage is eminently descriptive of the appearance of a great comet, and the occasion on which it is introduced adds to the intensity of the lurid imaginings and feelings of terror and dismay with which these objects have always been regarded. The comparison of the enraged Prince of Hell with one of those mysterious and fiery looking visitors to our skies was a grand conception of the poet's, and one worthy of the mighty combatant. Ophiuchus (the Serpent-bearer) is a large constellation which occupies a rather barren region of the heavens to the south of Hercules. It has a length of about forty degrees, and is represented by the figure of a man bearing a serpent in both hands. It is not easy to imagine why Milton should have assigned the comet

to this uninteresting constellation ; he may possibly
have seen one in this part of the sky, or his poetical
ear may have perceived that the expression 'Ophiuchus huge,' which has about it a ponderous rhythm,
was well adapted for the poetic description of a comet.

The only other allusion in the poem to a comet
is near its conclusion, when the Cherubim descend
to take possession of the Garden, prior to the removal
of Adam and Eve—

> High in front advanced,
> The brandished sword of God before them blazed,
> Fierce as a comet ; which with torrid heat,
> And vapour as the Lybian air adust
> Began to parch that temperate clime.—xii. 632–36.

FALLING STARS

On any clear night an observer can, by attentively watching the heavens, perceive a few of those
objects which become visible for a moment as a
streak of light and then vanish.   They are the result
of the combustion of small meteoric masses having
a celestial origin, and travelling with cosmical velocity, and which, in their headlong flight, become so
heated by contact with the Earth's atmosphere that
they are converted into glowing vapour.   This
vapour when it cools condenses into fine powder or
dust, and gradually descends upon the Earth's surface, where it can be detected.

Shooting stars become visible at a height varying between twenty and one hundred and thirty

miles, and their average velocity has been estimated at about thirty miles a second. Though casual falling stars can be seen at all times in every part of the heavens, yet there are certain periods at which they appear in large numbers, and have been observed to radiate from certain well-defined parts of the sky. When the radiant point is overhead, the falling stars spread out and resemble a parachute of fire; but when it is below the horizon, the stars ascend upwards like rockets into the sky. The radiant point is fixed among the stars, so that at the commencement of a shower it may be overhead, and before the termination of the display it may have travelled below the horizon. The radiant is usually named after the constellation in which it is observed.

The November meteors are called Leonids, because they radiate from a point in the constellation Leo; those in Taurus are called Taurids; in Perseus, Perseids; in Lyra, Lyraïds; and in Andromeda, Andromedes, because their radiant points are situated in those constellations.

The falling stars that have attracted most attention are those which appear on or about November 13. Every year at this period they can be seen in greater or less numbers, and on referring to numerous past records it has been ascertained that a magnificent display of those objects occurs every thirty-three years. The earliest historical allusion to this meteoric shower is by Theophanes, who wrote that in the year 472 A.D. the sky at Constantinople ap-

peared to be on fire with falling stars.   In the year
902 A.D. another remarkable display took place, and
from that time until 1833 twelve conspicuous displays are recorded as having occurred at recurring
intervals of thirty-three years.   The grandest display of this kind that was ever witnessed occurred
in 1833.   It was visible over nearly the whole of the
American continent, and, having commenced at midnight, lasted for four or five hours.   The falling stars
were so numerous that they appeared to rain upon
the Earth, and caused the utmost consternation and
terror among those who witnessed the phenomenon,
many persons having imagined that the end of the
world was at hand.   The regular recurrence of these
meteoric displays has been satisfactorily explained
by the assumption that round the Sun there travels
in an elliptical orbit with planetary velocity a vast
shoal of meteoric bodies some millions of miles in
length and several hundred thousand miles in
breadth.   The nearest point of their orbit to the
Sun coincides with the Earth's orbit, and the most
distant part extends beyond the orbit of Uranus.
These bodies accomplish a circuit of their orbit in
$33\frac{1}{4}$ years.   The Earth in her annual revolution
intersects the path of the meteors, and when this
occurs some falling stars can always be seen; but
when the intersection happens at the time the shoal
is passing, then there results a grand meteoric display.   Numerous other meteoric swarms travel in
orbital paths round the Sun.

Milton, in his poem, alludes to falling stars upon

two occasions. In describing the fall of Mulciber from Heaven he says :—

> from morn
> To noon he fell, from noon to dewy eve,
> A summer's day ; and with the setting sun
> Dropt from the zenith like a falling star,
> On Lemnos the Ægaean isle.—i. 742–46.

The rapid flight of the archangel Uriel from the Sun to the Earth is described in the following lines :—

> Thither came Uriel, gliding through the even
> On a sunbeam, swift as a shooting star
> In autumn thwarts the night, when vapours fired
> Impress the air, and shows the mariner
> From what point of his compass to beware
> Impetuous winds.—iv. 555–60.

Milton mentions the season of the year in which those stars are most frequently seen, and refers to an ancient belief by which they were regarded as the precursors of stormy weather. A translation from Virgil contains a similar allusion to them—

> Oft shalt thou see ere brooding storms arise,
> Star after star glide headlong down the skies.

The standard borne by the Cherub Azazel is described as having—

> Shone like a meteor streaming to the wind.—i. 537.

CHAPTER IX

MILTON'S IMAGINATIVE AND DESCRIPTIVE ASTRONOMY

THE theme chosen by Milton for his great epic, viz. the Fall of Man and his expulsion from Paradise— perhaps the most momentous incident in the history of the human race—was one worthy of the genius of a great poet and in the treatment of which Milton has been sublimely successful. The newly created Earth; the untainted loveliness of the Paradise in which our first parents dwelt during their inno- cence; their temptation; their fall and removal from the happy garden, furnished a theme which afforded him an opportunity for the display of his unrivalled poetic genius.

Though the chief interest in the poem is centred in the Garden of Eden and its occupants, yet Milton was enabled, by the comprehensive manner in which he treated his subject, to intro- duce into his work a cosmology which embraced not only the system to which our globe belongs, but the entire starry heavens by which we are sur- rounded. But the universality of his genius did not rest here. In the utterance of his sacred song he soared beyond the starry sphere, describing himself as wrapt above the pole—the starry pole—up to the Empyrean, or Heaven of Heavens, the ineffable abode of the Deity and the blissful habitation of angelic beings who, in adoration and worship, surround the throne of the Most High.

Descending to that nether world at the opposite pole of the universe, in the lowest depth of Chaos, the place prepared by Eternal Justice for the rebellious, he unfolds to our horror-stricken gaze the terrors of this infernal region ; its fiery deluge of ever-burning sulphur ; its ' regions of sorrow ; ' its ' doleful shades '—the unhappy abode of fallen angels who ' in floods and whirlwinds of tempestuous fire,' alternated by exposure to unendurable cold and icy torment,experience the direful consequences of their apostacy.

Milton's ' Paradise Lost ' may be regarded as the loftiest intellectual effort in the whole range of literature. In it we find all that was known of science, philosophy, and theology. The theme, founded upon a Bible narrative, itself written under divine inspiration, embraces the entire system of Christian doctrine as revealed in the Scriptures, and many of the noblest passages in the sacred volume are introduced into the poem expressed in the lofty utterance of flowing and harmonious verse. The choicest classical writings of Greek and Latin authors ; the mythological and traditional beliefs of ancient nations ; historical incidents of valour and renown and all that was great and good in the annals of mankind were laid under contribution by

Milton in the illustration and embellishment of his poem.

In order to obtain a basis or foundation upon which to construct his great epic, Milton found it necessary to localise the regions of space in which the principal events mentioned in his poem are described as having occurred. The unfathomable abyss of space may be regarded as an uncircumscribed sphere boundless on all sides round, and so far as we can comprehend of infinite extent. This sphere Milton divided into two hemispheres—an upper and a lower. The upper was called Heaven, or the Empyrean—a glorified region of boundless dimensions; the lower hemisphere embraced Chaos—a dark, fathomless abyss in which the elements of matter existed in a state of perpetual tumult and wild uproar. The occurrence of a rebellion in Heaven necessitated a further division of the sphere. The revolt, headed by Lucifer, one of the highest archangels, afterwards known as Satan, who drew after him a third of the angelic host, contested the supremacy of Heaven with Michael and the angels which kept their loyalty. After two days' battle—

> Him the Almighty Power
> Hurled headlong flaming from the ethereal sky,
> With hideous ruin and combustion, down
> To bottomless perdition; there to dwell
> In adamantine chains and penal fire.—i. 44–48.

Having been precipitated over the crystal wall of Heaven into the deep abyss, Milton says :—

Nine days they fell; confounded Chaos roared,
And felt tenfold confusion in their fall
Through his wild Anarchy; so huge a rout
Encumbered him with ruin. Hell at last,
Yawning, received them whole, and on them closed.

vi. 871–75.

Hell, Milton locates in the lowest depth of Chaos, a region cut off from the body of Chaos, through which the expelled angels fell for nine days before reaching their destined habitation. There are now three divisions of space: HEAVEN, CHAOS, and HELL. But a fourth is required to enable Milton to complete his scheme for the delineation of his poem. The Earth and starry universe were not as yet called into existence, but after the overthrow of the rebellious angels, God, by circumscribing a portion of Chaos situated immediately underneath the Empyrean, created the Mundane Universe, or the ' Heavens and the Earth.' [1] This new universe He reclaimed from Chaos, and with the embryo elements of matter—

His dark materials to create new worlds.—ii. 916,

He formed the Earth and all the countless shining orbs visible overhead, and the myriads more which the telescope reveals, scattered in apparently endless profusion over the circular immensity of space. It is this new universe—the Earth and Starry Heavens—that claims our chief attention, and in the delineation of Milton's imaginative and descriptive powers it is to this latest manifestation of Divine

[1] See diagram, chap. iii. p. 96.

wisdom and might that our remarks shall princi-
pally apply. After the expulsion of the rebel angels
from Heaven, God sent His Son, the Messiah, to
create the new universe—a work of omnipotence
described by Milton in a manner worthy of so magnificent a display of almighty power—

> Meanwhile the Son
> On his great expedition now appeared,
> Girt with omnipotence, with radiance crowned
> Of majesty divine : sapience and love
> Immense ; and all his Father in Him shone.
> About his chariot numberless were poured
> Cherub and Seraph, Potentates and Thrones,
> And Virtues, winged Spirits, and chariots winged
> From the armoury of God, where stand of old
> Myriads, between two brazen mountains lodged
> Against a solemn day, harnessed at hand,
> Celestial equipage ; and now came forth
> Spontaneous, for within them Spirit lived,
> Attendant on their Lord.   Heaven opened wide
> Her ever-during gates, harmonious sound !
> On golden hinges moving, to let forth
> The King of Glory, in his powerful Word
> And Spirit, coming to create new worlds.
> On Heavenly ground they stood, and from the shore
> They viewed the vast immeasurable abyss
> Outrageous as a sea, dark, wasteful, wild,
> Up from the bottom turned by furious winds
> And surging waves, as mountains to assault
> Heaven's highth, and with the centre mix the pole.
> ' Silence, ye troubled Waves, and thou Deep, peace ! '
> Said then the omnific Word : ' your discord end ! '
> Nor stayed ; but on the wings of Cherubim
> Uplifted, in paternal glory rode
> Far into Chaos, and the World unborn ;
> For Chaos heard his voice.   Him all his train
> Followed in bright procession, to behold
> Creation, and the wonders of his might.

Then stayed the fervid wheels, and in his hand
He took the golden compasses, prepared
In God's eternal store, to circumscribe
This Universe, and all created things.
One foot he centred, and the other turned
Round through the vast profundity obscure ;
And said, ' Thus far extend, thus far thy bounds ;
This be thy just circumference, O World ! '
Thus God the Heaven created, thus the Earth,
Matter unformed and void. Darkness profound
Covered the abyss ; but on the watery calm
His brooding wings the Spirit of God outspread,
And vital virtue infused, and vital warmth,
Throughout the fluid mass ; but downward purged
The black, tartareous, cold, infernal dregs,
Adverse to life ; then founded, then conglobed
Like things to like ; the rest to several place
Disparted, and between spun out the Air ;
And Earth self balanced on her centre hung.

vii. 192–242.

Milton begins his narrative of the Creation by describing the progress of the Deity on His great expedition, accompanied by hosts of angels and surrounded with all the solemn pomp and splendour of Heaven. The brilliant throng having passed through Heaven's gates, which opened wide their portals, they beheld in front of them the dark abyss of Chaos—a tempest-tossed sea of warring elements upturned in wild confusion. At God's instant command silence and peace reigned over the deep, and tranquil calm succeeded noisy discord. Then on the wings of Cherubim He rode far into Chaos, and with His golden compasses decreed the dimensions of the universe by circumscribing the vast vacuity of space. Into the elements which hasted to their several places, His Spirit infused vital warmth and caused the formless mass of matter to assume the figure of a sphere, and thus the Earth poised on her axis unsupported, and in darkness shrouded hung suspended in space. The placing of the golden compasses in the hands of the Creator, with which He measured out the heavens, is a noble conception on the part of Milton, and one most appropriate, since the construction of the universe is based upon the principles of geometrical science.

'Let there be Light!' said God; and forthwith Light
Ethereal, first of things, quintessence pure,
Sprung from the Deep; and from her native east
To journey through the aëry gloom began,
Sphered in a radiant cloud; for yet the Sun
Was not; she in a cloudy tabernacle
Sojourned the while. God saw the light was good;
And light from darkness by the hemisphere
Divided; light the day, and darkness night
He named. Thus was the first day even and morn:
Nor passed uncelebrated, nor unsung
By the celestial quires, when orient light
Exhaling first from darkness they beheld;
Birthday of Heaven and Earth; with joy and shout
The hollow universal orb they filled,
And touched their golden harps, and hymning praised
God and his works: Creator Him they sung,
Both when first evening was, and when first morn.

vii. 243-60.

The appearance of Light, which sprung into existence at the fiat of the Creator, was the next great event witnessed by beholding angels—birthday of Heaven and Earth, first morning and first evening, which the celestial choirs celebrated with praise and shouts of joy. The creation of the firmament was the great work of the second day.

> Again God said, ' Let there be firmament
> Amid the waters, and let it divide
> The waters from the waters ! ' And God made
> The firmament, expanse of liquid, pure,
> Transparent, elemental air, diffused
> In circuit to the uttermost convex
> Of this great round—partition firm and sure,
> The waters underneath from those above
> Dividing; for as the Earth, so He the World
> Built on circumfluous waters calm, in wide
> Crystalline ocean, and the loud misrule
> Of Chaos far removed, lest fierce extremes
> Contiguous might distemper the whole frame :
> And Heaven he named the Firmament. So even
> And morning chorus sung the second day.—vii. 261–275.

After describing the gathering of the waters off the face of the globe into seas, causing the dry land to appear, which at the word of God became clothed with vegetation, rendering the Earth a habitable abode, Milton proceeds to describe the creation of the heavenly bodies—

> Again the Almighty spake : ' Let there be Lights
> High in the expanse of Heaven, to divide
> The day from night ; and let them be for signs,
> For seasons, and for days, and circling years ;
> And let them be for lights, as I ordain
> Their office in the firmament of Heaven,
> To give light on the Earth ! ' and it was so.
> And God made two great Lights, great for their use
> To Man, the greater to have rule by day,
> The less by night, altern ; and made the Stars,
> And set them in the firmament of Heaven
> To illuminate the Earth, and rule the day
> In their vicissitude, and rule the night,
> And light from darkness to divide. God saw,

Surveying his great work, that it was good :
For, of celestial bodies, first, the Sun,
A mighty sphere He framed, unlightsome first,
Though of ethereal mould ; then formed the Moon
Globose, and every magnitude of Stars,
And sowed with stars the Heaven thick as a field.
Of light by far the greater part he took,
Transplanted from her cloudy shrine, and placed
In the Sun's orb, made porous to receive
And drink the liquid light ; firm to retain
Her gathered beams, great palace now of Light.
Hither, as to their fountain, other stars
Repairing, in their golden urns draw light,
And hence the morning planet gilds her horns ;
By tincture or reflection they augment
Their small peculiar, though, from human sight
So far remote, with diminution seen.
First in his east the glorious lamp was seen,
Regent of day, and all the horizon round
Invested with bright rays, jocund to run
His longitude through Heaven's high road ; the grey
Dawn, and the Pleiades before him danced,
Shedding sweet influence.   Less bright the Moon,
But opposite in levelled west was set
His mirror, with full face borrowing her light
From him ; for other light she needed none
In that aspect, and still that distance keeps
Till night ; then in the east her turn she shines,
Revolved on Heaven's great axle, and her reign
With thousand lesser lights dividual holds,
With thousand thousand stars that then appeared
Spangling the hemisphere.   Then first adorned
With their bright luminaries, that set and rose,
Glad evening and glad morn crowned the fourth day.

vii. 339–86.

The first creation was Light, and Milton,
according to Scriptural testimony, ascribes its
origin to the bidding of the Creator. 'God said, Let
there be light ; and there was light !' The Sun

he describes as a mighty sphere, but at first non-luminous. There was light, but no sun. The reason usually given in explanation of this pheno-menon is, that the heavenly bodies were created at the same time as the Earth, but were rendered invisible by a canopy of vapour and cloud which enveloped the newly-formed globe ; and that after-wards, when it dispersed, they appeared in the firmament, shining in all their pristine splendour. Milton does not, however, adhere to this view of things, but says that light for the first three days sojourned in a cloudy shrine or tabernacle, and was afterwards transplanted in the Sun, which became a great palace of light.

He expresses himself in a somewhat similar manner in Book III., which opens with an address to Light—one of the most beautiful passages in the poem, in which he alludes to his blindness when expressing his thoughts and sentiments with regard to this ethereal medium, which conveys to us the pleasurable sensation of vision—

Hail, holy Light ! offspring of Heaven first-born !
Or of the Eternal co-eternal beam,
May I express thee unblamed? since God is light,
And never but in unapproached light
Dwelt from eternity—dwelt then in thee,
Bright effluence of bright essence increate !
Or hear'st thou rather, pure Ethereal stream,
Whose fountain who shall tell? Before the Sun,
Before the Heavens thou wert, and at the voice
Of God, as with a mantle, didst invest
The rising world of waters dark and deep,
Won from the void and formless Infinite.—iii. 1–12.

The Sun having become a lucent orb, Milton
poetically describes how the planets repair to him
as to a fountain, and in their golden urns draw
light; and how the morning planet Venus gilds
her horns illumined by his rays. The poet associates joyous ideas with the new-born universe.
The Sun, now the glorious regent of day, begins
his journey in the east, lighting up the horizon
with his beams; whilst before him danced the grey
dawn, and the Pleiades shedding sweet influences.
There existed an ancient belief that the Earth was
created in the spring, and in April the Sun is in
the zodiacal constellation Taurus, in which are
also situated the Pleiades; they rise a little before
the orb, and precede him in his path through the
heavens. The stars of this group have always
been regarded with a peculiar sacredness, and their
rays, mingling with those of the Sun, were believed
to shed sweet influences upon the Earth. The
Moon, less bright, with borrowed light, in her turn
shines in the east, and, with the thousand thousand
luminaries that spangle the firmament, reigns over
the night.

We learn in Book III. that the archangel Uriel,
who was beguiled by Satan, witnessed the Creation, and described how the heavenly bodies were
brought into existence, he having perceived what
we should call the gaseous elements of matter
rolled into whorls and vortices which became condensed into suns and systems of worlds. This
mighty angel says :—

> I saw when, at his word the formless mass,
> This World's material mould, came to a heap :
> Confusion heard his voice, and wild Uproar
> Stood ruled, stood vast Infinitude confined ;
> Till at his second bidding darkness fled,
> Light shone, and order from disorder sprung.
> Swift to their several quarters hasted then
> The cumbrous elements, Earth, Flood, Air, Fire ;
> And this ethereal quintessence of Heaven
> Flew upward, spirited with various forms,
> That rolled orbicular, and turned to stars
> Numberless, as thou seest, and how they move ;
> Each had his place appointed, each his course ;
> The rest in circuit walls this Universe.—iii. 708-21.

In his sublime description of the Creation Milton has adhered with marked fidelity to the Mosaic version, as narrated in the first two chapters of Genesis, when God, by specific acts in certain stated periods of time, created the visible universe and all that it contains.

The successive acts of creation are described in words almost identical with those of Scripture, embellished and adorned with all the wealth of expression which our language is capable of affording. The several scenes presented to the imagination, and witnessed by hosts of admiring angels as each portion of the magnificent work was accomplished, are full of a grandeur and majesty worthy of the loftiest conceivable effort of Divine power and might.

The return of the Creator after the completion of His great work is described by Milton in a manner worthy of the progress of Deity through the celestial regions. The whole creation rang with jubilant delight, and the bright throng which witnessed the wonders of His might followed Him with acclamation, ascending by the glorified path of the Milky Way up to His high abode—the Heaven of Heavens—

> Here finished He, and all that He had made
> Viewed, and behold! all was entirely good.
> So even and morn accomplished the sixth day:
> Yet not till the Creator from his work
> Desisting, though unwearied, up returned,
> Up to the Heaven of Heavens, His high abode,
> Thence to behold this new created World,
> The addition of his empire, how it showed
> In prospect from His throne, how good, how fair,
> Answering his great idea.   Up He rode,
> Followed with acclamation, and the sound
> Symphonious of ten thousand harps, that tuned
> Angelic harmonies: The Earth, the Air
> Resounded (thou remember'st, for thou heard'st)
> The Heavens and all the constellations rung,
> The planets in their stations listening stood,
> While the bright pomp ascended jubilant.
> ' Open ye everlasting gates!' they sung;
> ' Open ye Heavens! your living doors; let in
> The great Creator, from his work returned
> Magnificent, his six days' work, a World;
> Open, and henceforth oft; for God will deign
> To visit oft the dwellings of just men,
> Delighted; and with frequent intercourse
> Thither will send his winged messengers
> On errands of supernal grace.'   So sung
> The glorious train ascending: He through Heaven,
> That opened wide her blazing portals, led
> To God's eternal house direct the way—
> A broad and ample road, whose dust is gold,
> And pavement stars, as stars to thee appear
> Seen in the Galaxy, that Milky Way
> Which nightly as a circling zone thou seest
> Powdered with stars.—vii. 548-81.

Milton, throughout his description of the Creation, sustains with lofty eloquence his sublime conception of this latest display of almighty power; and invests with becoming majesty all the acts of the Creator, who, when He finished His great work, saw that all was entirely good.

Shortly after the creation of the new universe, Satan, having escaped from Hell, plunged into the abyss of Chaos, and, after a long and arduous journey upwards, in which he had to fight his way through the surging elements that raged around him like a tempestuous sea, he reached the upper confines of this region where less confusion prevailed, and where a glimmering dawn of light penetrated its darkness and gloom, indicating that the limit of the empire of Chaos and ancient Night had been reached by the adventurous fiend. Pursuing his way with greater ease, he leisurely beholds the sight which is opening to his eyes—a sight rendered more glorious by his long sojourn in darkness. He sees :—

> Far off the empyreal Heaven, extended wide
> In circuit, undetermined square or round,
> With opal towers and battlements adorned
> Of living sapphire, once his native seat,
> And, fast by, hanging in a golden chain,
> This pendent World, in bigness as a star
> Of smallest magnitude close by the Moon.
>
> ii. 1047–53.

He gazes upon his native Heaven where once he dwelt, and observes the pendent world in quest of which he journeyed hither—hung by a golden

chain from the Empyrean and no larger than a star
of the smallest magnitude when close by the Moon.
In this passage Milton does not allude to the Earth,
which was invisible, but to the entire starry heavens
—the newly created universe reclaimed from Chaos,
which, when contrasted with the Empyrean, appeared in size no larger than the minutest star
when compared with the full moon. Pursuing his
journey, the new universe as it is approached ex-
pands into a globe of vast dimensions ; its convex
surface—round which the chaotic elements in
stormy aspect lowered—seemed a boundless continent, dark, desolate, and starless, except on the
side next to the wall of Heaven, which though far-
distant afforded it some illumination by its reflected
light.   Satan, having alighted on this convex shell
which enclosed the universe, wandered long over
its bleak and dismal surface, until his attention was
attracted by a gleam of light which appeared
through an opening at its zenith right underneath
the Empyrean. Thither he directed his steps, and
perceived a structure resembling a staircase, or
ladder, which formed the only means of communication between Heaven and the new creation, and
upon which angels descended and ascended—

> Far distant he descries,
> Ascending by degrees magnificent
> Up to the wall of Heaven, a structure high ;
> At top whereof, but far more rich, appeared
> The work as of a kingly palace gate,
> With frontispiece of diamond and gold
> Embellished ; thick with sparkling orient gems

The portal shone, inimitable on Earth
By model, or by shading pencil drawn.
The stairs were such as whereon Jacob saw
Angels ascending and descending, bands
Of Guardians bright, when he from Esau fled
To Padan Aram, in the field of Luz
Dreaming by night under the open sky,
And waking cried, ' *This is the gate of Heaven.*'

iii. 501–15.

Sometimes this mysterious structure was drawn up to Heaven and invisible. At the time that Satan reached the opening, the stairs were lowered, and standing at their base he looked down with wonder upon the entire starry universe —

Such wonder seized, though after Heaven seen,
The Spirit malign, but much more envy seized,
At sight of all this World beheld so fair,
Round he surveys (and well might, where he stood
So high above the circling canopy
Of night's extended shade) from eastern point
Of Libra to the fleecy star that bears
Andromeda far off Atlantic seas
Beyond the horizon ; then from pole to pole
He views in breadth, and without longer pause,
Down right into the World's first region throws
His flight precipitant, and winds with ease
Through the pure marble air his oblique way
Amongst innumerable stars, that shone
Stars distant, but nigh hand seemed other worlds,
Or other worlds they seemed, or happy isles,
Like those Hesperian Gardens famed of old,
Fortunate fields, and groves, and flowery vales ;
Thrice happy isles !   But who dwelt happy there
He staid not to inquire : above them all
The golden Sun, in splendour likest Heaven
Allured his eye : thither his course he bends
Through the calm firmament, (but up or down

Y

By centre or eccentric hard to tell
Or longitude) where the great luminary,
Aloof the vulgar constellations thick,
That from his lordly eye keep distance due,
Dispenses light from far.   They, as they move
Their starry dance in numbers that compute
Days, months, and years, towards his all-cheering lamp
Turn swift their various motions, or are turned
By his magnetic beam, that gently warms
The Universe, and to each inward part
With gentle penetration, though unseen,
Shoots invisible virtue even to the Deep;
So wondrously was set his station bright.

iii. 552–87.

The Ptolemaic cosmology having been adopted
by Milton in the elaboration of his poem, he
describes the universe in conformity with the doctrines associated with this form of astronomical
belief.   To each of the first seven spheres which
revolved round the steadfast Earth there was
attached a heavenly body ; the eighth sphere embraced all the fixed stars, a countless multitude ;
the ninth the crystalline ; and enclosing all the
other spheres as if in a shell was the tenth sphere,
or Primum Mobile, which in its diurnal revolution
carried round with it all the other spheres.   The
nine inner spheres were transparent, but the tenth
was an opaque solid shell-like structure, which
enclosed the new universe and constituted the
boundary between it and Chaos underneath and
the Empyrean above.   It was on the surface of
this sphere that Satan wandered until he discovered the opening at its zenith, where, by means

of a staircase or ladder, communication was maintained with the Empyrean. Standing on the lower steps of this structure he paused for a moment to look down into the glorious universe which lay beneath him—

> another Heaven
> From Heaven-gate not far, founded in view
> On the clear hyaline the glassy sea.—vii. 617–19.

He beholds it in all its dimensions, from pole to pole, and longitudinally from Libra to Aries, then without hesitation precipitates himself down into the world's first region, and winds his way with ease among the fixed stars. Around him he sees innumerable shining worlds, sparkling and glittering in endless profusion over the circumscribed immensity of space—mighty constellations that shone from afar; clustering aggregations of stars; floating islands of light; twinkling systems rising out of depths still more profound, and a zone luminous with the light of myriads of lucid orbs verging on the confines of the universe. All these worlds the fiend passed unheeded, nor stayed he to inquire who dwelt happy there. In splendour above them all the Sun attracted his attention and, directing his course towards the great luminary of our system, he alights on the surface of the orb.

Milton now makes a digression in order to describe what Satan observed in the Sun after having landed there. The poet embraces an opportunity for exercising his imaginative and descriptive powers by giving an ideal description of what,

judging from the appearance of the orb, might be
the natural condition of things existing on his surface—

> There lands the Fiend, a spot like which perhaps
> Astronomer in the Sun's lucent orb
> Through his glazed optic tube, yet never saw.
> The place he found beyond expression bright,
> Compared with aught on Earth, metal or stone ;
> Not all parts like, but all alike informed
> With radiant light, as glowing iron with fire ;
> If metal, part seemed gold, part silver clear ;
> If stone, carbuncle most or chrysolite,
> Ruby or topaz, to the twelve that shone
> In Aaron's breastplate, and a stone besides,
> Imagined rather oft than elsewhere seen ;
> That stone, or like to that, which here below
> Philosophers in vain so long have sought,
> In vain, though by their powerful art they bind
> Volatile Hermes, and call up unbound
> In various shapes old Proteus from the sea,
> Drained through a limbec to his native form.
> What wonder then if fields and regions here
> Breathe forth elixir pure, and rivers run
> Potable gold, when, with one virtuous touch,
> The arch-chemic Sun, so far from us remote,
> Produces, with terrestrial humour mixed,
> Here in the dark so many precious things
> Of colour glorious, and effect so rare ?
> Here matter new to gaze the Devil met
> Undazzled ; far and wide his eye commands ;
> For sight no obstacle found here, nor shade,
> But all sunshine, as when his beams at noon
> Culminate from the equator, as they now
> Shot upward still direct, whence no way round
> Shadow from body opaque can fall ; and the air,
> Nowhere so clear sharpened his visual ray
> To objects distant far, whereby he soon
> Saw within here a glorious Angel stand.—iii. 588–622.

The physical structure of the interior of the

Sun is unknown; all that we see of the orb is the photosphere—the dazzling luminous envelope which indicates to the eye the boundary of the solar disc, and which is the source of light and heat. Milton, in his imaginative and beautifully poetical description of the Sun, is not more fanciful in his conception of the nature of the refulgent orb than a renowned astronomer (Sir William Herschel) who writes in the following strain: 'A cool, dark, solid globe, its surface diversified with mountains and valleys, clothed in luxuriant vegetation and richly stored with inhabitants, protected by a heavy cloud-canopy from the intolerable glare of the upper luminous region, where the dazzling coruscations of a solar aurora some thousands of miles in depth evolved the stores of light and heat which vivify our world.' Satan, disguised as a cherub, makes himself known to Uriel, Regent of the Sun. The upright Seraph in response to his request directs him to the Earth, the abode of Man—

> Look downward on that Globe, whose hither side
> With light from hence, though but reflected, shines,
> That place is Earth, the seat of Man ; that light
> His day, which else, as the other hemisphere,
> Night would invade; but there neighbouring Moon
> (So call that opposite fair star) her aid
> Timely interposes, and her monthly round
> Still ending, still renewing, through mid-Heaven,
> With borrowed light her countenance triform
> Hence fills and empties, to enlighten the Earth,
> And in her pale dominion checks the night.
>
> iii. 722–32.

It would be impossible not to feel impressed

with the accuracy and comprehensiveness of
Milton's astronomical knowledge ; and how he has
united in charming poetic expression the dry details
of science with the divine inspiration of the
heavenly muse. The distinctive appearances of the
Sun, Moon, planets, and stars ; their functional
importance as regards this terrestrial sphere ; the
splendour and lustre peculiar to each ; and the
glory displayed in the entire created heavens, are
portrayed with a skill indicative of a masterly
knowledge of the science of astronomy.

> DESCEND from Heaven, Urania, by that name
> If rightly thou art called, whose voice divine
> Following, above the Olympian hill I soar,
> Above the flight of Pegasean wing !
> The meaning, not the name, I call ; for thou
> Nor of the Muses nine, nor on the top
> Of old Olympus dwell'st ; but heavenly-born,
> Before the hills appeared or fountain flowed,
> Thou with Eternal Wisdom didst converse,
> Wisdom thy sister, and with her didst play
> In presence of the Almighty Father, pleased
> With thy celestial song.   Up led by thee,
> Into the Heaven of Heavens I have presumed,
> An earthly guest, and drawn empyreal air,
> Thy tempering.   With like safety guided down,
> Return me to my native element ;
> Lest, from this flying steed unreined, (as once
> Belerophon, though from a lower clime)
> Dismounted, on the Aleian field I fall,
> Erroneous there to wander, and forlorn.
> Half yet remains unsung, but narrower bound
> Within the visible diurnal sphere.
> Standing on Earth, not rapt above the pole,
> More safe I sing with mortal voice, unchanged
> To hoarse or mute, though fallen on evil days,

On evil days though fallen, and evil tongues,
In darkness, and with dangers compassed round,
And solitude ; yet not alone, while thou
Visit'st my slumbers nightly, or when morn
Purples the east. Still govern thou my song,
Urania, and fit audience find though few.—vii. 1-32.

The Muses were Greek mythological divinities who possessed the power of inspiring song, and were the patrons of poets and musicians. According to Hesiod they were nine in number and presided over the arts. Urania was the Goddess of Astronomy, and Calliope the Goddess of Epic Poetry. They are described as the daughters of Zeus, and Homer alludes to them as the goddesses of song who dwelt on the summit of Mount Olympus. They were the companions of Apollo, and accompanied with song his playing on the lyre at the banquets of the Immortals. Milton does not invoke the mythological goddess, but Urania the Heavenly Muse, whose aid he also implores at the commencement of his poem prior to his flight above the Aonian Mount. Under her divine guidance he ascended to the Heaven of Heavens and breathed empyreal air, her tempering ; in like manner he requests her to lead him down to his native element lest he should meet with a fate similar to what befell Bellerophon. Half his task he has completed, the other half, confined to narrower bounds within the visible diurnal sphere, remains unsung, and in its fulfilment he still implores his celestial patroness to govern his song.

The natural phenomena which occur as a consequence of the motions of the heavenly bodies and the diurnal rotation of the Earth on her axis, are accompanied by agreeable alternations in the aspect of nature with which every one is familiar. The rosy footsteps of morn ; the solar splendour of noon-day ; the fading hues of even ; and night with her jewelled courts and streams of molten stars, have been sung with rapturous admiration by poets of every nation and in every age. They, as ardent lovers of nature, have described in choicest language the pleasing vicissitudes brought about by the real and apparent motions of the celestial orbs.

In this respect Milton is unsurpassed by any poet in ancient or in modern times. The occasions on which he describes the heavenly bodies, or alludes to them in association with other phenomena, testify to the felicity of his thoughts and to the greatness of his poetic genius. Surely no poet has ever given us a lovelier description of evening, or has added more to its exquisite beauty by his allusion to the celestial orbs, than Milton when he describes the first evening in Paradise—

> Now came still Evening on, and Twilight gray
> Had in her sober livery all things clad ;
> Silence accompanied ; for beast and bird,
> They to their grassy couch, these to their nests
> Were slunk, all but the wakeful nightingale.
> She all night long her amorous descant sung ;
> Silence was pleased. Now glowed the firmament
> With living sapphires : Hesperus that led

The starry host, rode brightest, till the Moon,
Rising in clouded majesty, at length
Apparent queen, unveiled her peerless light,
And o'er the dark her silver mantle threw.

iv. 598–609.

In the avowal of her conjugal love, Eve, with charming expression, associates the orbs of the firmament with the delightful appearances of nature which presented themselves to her observation after she awoke to the consciousness of intelligent existence.

Sweet is the breath of Morn, her rising sweet,
With charm of earliest birds: pleasant the Sun,
When first on this delightful land he spreads
His orient beams, on herb, tree, fruit, and flower,
Glistering with dew; fragrant the fertile Earth
After soft showers; and sweet the coming on
Of grateful Evening mild; then silent Night,
With this her solemn bird, and this fair Moon,
And these the gems of Heaven, her starry train :
But neither breath of Morn, when she ascends
With charm of earliest birds; nor rising Sun
On this delightful land; nor herb, fruit, flower,
Glistering with dew; nor fragrance after showers;
Nor grateful Evening mild; nor silent Night,
With this her solemn bird; nor walk by Moon,
Or glittering star-light, without thee is sweet.
But wherefore all night long shine these ? for whom
This glorious sight, when sleep hath shut all eyes ?

iv. 641–58.

One of the charms of Milton's verse is the devoutly poetical sentiment which pervades it. His thoughts, though serious, are not austere or gloomy, and it is in his loftiest musings that his reverence becomes most apparent. This feeling is conspicuous in Adam's reply to the inquiry addressed to him by Eve—

> Daughter of God and Man, accomplished Eve,
> These have their course to finish round the Earth
> By morrow evening, and from land to land
> In order, though to nations yet unborn,
> Ministering light prepared, they set and rise ;
> Lest total Darkness should by night regain
> Her old possession, and extinguish life
> In Nature and all things ; which these soft fires
> Not only enlighten, but with kindly heat
> Of various influence foment and warm,
> Temper or nourish, or in part shed down
> Their stellar virtue on all kinds that grow
> On Earth, made hereby apter to receive
> Perfection from the Sun's more potent ray.
> These, then, though unbeheld in deep of night,
> Shine not in vain ; nor think, though men were none,
> That Heaven would want spectators, God want praise :
> Millions of spiritual creatures walk the Earth
> Unseen, both when we wake, and when we sleep :
> All these with ceaseless praise his works behold
> Both day and night.   How often from the steep
> Of echoing hill or thicket, have we heard
> Celestial voices to the midnight air,
> Sole, or responsive each to other's note
> Singing their Great Creator !   Oft in bands
> While they keep watch, or nightly rounding walk,
> With heavenly touch of instrumental sounds
> In full harmonic number joined, their songs
> Divide the night, and lift our thoughts to Heaven.
>
> iv. 660–88.

The Morning Hymn of Praise which Adam and Eve offer up in concert to their Maker contains their loftiest thoughts and most reverent sentiments, expressed in melodiously flowing verse.   In their solemn invocations they call upon the orbs of the firmament to join in praising and extolling the Creator, and in their devout enthusiasm and adoration address by name those that are most conspicuous. Hesperus, 'fairest of stars,' is asked to praise Him in her sphere. The Sun, great image of his Maker, is told to acknowledge Him his greater, and to sound His praise in his eternal course. The Moon, the fixed stars, and the planets are called upon to resound the praise of the Creator, whose glory is declared in the Heavens—

> Fairest of Stars, last in the train of night,
> If better thou belong not to the dawn,
> Sure pledge of day, that crown'st the smiling morn
> With thy bright circlet, praise Him in thy sphere
> While day arises, that sweet hour of prime.
> Thou Sun, of this great world both eye and soul,
> Acknowledge Him thy greater; sound his praise
> In thy eternal course, both when thou climb'st,
> And when high noon hast gained, and when thou fall'st.
> Moon, that now meet'st the orient Sun, now fliest
> With the fixed stars, fixed in their orb that flies;
> And ye five other wandering Fires, that move
> In mystic dance, not without song, resound
> His praise, who out of darkness called up Light.
>
> v. 166–79.

Milton's conception of celestial distances, and of the vast regions of interstellar space, is finely described in the following lines :—

> Down thither prone in flight
> He speeds, and through the vast ethereal sky
> Sails between worlds and worlds, with steady wing
> Now on the polar winds; then with quick fan
> Winnows the buxom air, till, within soar
> Of towering eagles.—v. 266–71.

As in their morning, so in their evening devotions, our first parents never fail to introduce a reference to the celestial orbs as indicating the power and goodness of the Creator, made manifest in the beauty and greatness of His works—

> Thus, at their shady lodge arrived, both stood,
> Both turned, and under open sky adored
> The God that made both Sky, Air, Earth and Heaven
> Which they beheld ; the Moon's resplendent globe,
> And starry pole.—iv. 720–24.

The numerous extracts contained in this volume impress upon one's mind how largely astronomy enters into the composition of ' Paradise Lost,' and of how much assistance the knowledge of this science was to Milton in the elaboration of his poem. Indeed, it would be hard to imagine how such a work could have been written except by a poet who possessed a proficient and comprehensive knowledge of astronomy. The chief characteristic of Milton's poetry is its sublimity, which is the natural outcome of the magnificence of his conceptions and of his own pure imaginative genius. Among all the fields of literature, science, and philosophy explored by him, he found none more congenial to his tastes, or that afforded his imagination more freedom for its loftiest flights, than the sublimest of sciences—astronomy. Whether we admire most the accuracy of his astronomical knowledge, or the wonderful creations of his poetic fancy, or his beautiful descriptions of the celestial orbs, it is apparent that in this domain of science, as a poet, he stands alone and without a rival. In his choice of the Ptolemaic cosmology Milton adopted a system with which he had been familiar from his youth—the same which his favourite poet Dante introduced into his poem, 'The Divina Commedia,' and which was well adapted for poetic description. The picturesque conception of ten revolving spheres, carrying along with them the orbs assigned to each, which, by their revolution round the steadfast Earth, brought about with unfailing regularity the successive alternation of day and night, and in every twenty-four hours exhibited the pleasing vicissitudes of dawn, of sunshine, of twilight, and of darkness, relieved by the soft effulgence of the nocturnal sky, afforded Milton a favourable basis upon which to construct a cosmical epic. The Copernican theory—with which he was equally conversant, and in the accuracy and truthfulness of which he believed—though less complicated than the Ptolemaic in its details, did not possess the same attractiveness for poetic description that belonged to the older system. According to this theory there is, surrounding us on all sides, a boundless uncircumscribed ocean of space, to which it is impossible to assign any conceivable limit ; in every effort to comprehend its dimensions or fathom its depths, the mind recoils upon itself, baffled and discomfited, with a conscious feeling that there can be no nearer approach to the end when end there is none that can be conceived of. Interspersed throughout the regions of

this azure vast of space is the stellar universe,
which to our comprehension is as infinite as the
abyss in which it exists.   The solar system, though
of magnificent dimensions, is but a unit in the
astronomical whole, in which are embraced millions
of other similar units—other solar systems, perhaps
differing in construction from that of ours, with
billions of miles of interstellar space intervening
between each ; yet so vast are the dimensions of
the celestial sphere that those distances when
measured upon it sink into utter insignificance. As
the receding depths of space are penetrated by
powerful telescopes, they are found to be pervaded
with stars and starry archipelagoes, distributed in
profusion over the circular immensity and extending
away into abysmal depths, beyond the reach of
visibility by any optical means which we possess.
To the universe there is no known end—nowhere in
imagination can its boundary be reached ! This
bewildering conception of the cosmos did not
trouble the minds of pre-Copernican thinkers.
They regarded the steadfast Earth as the most
important body in the universe ; nor were the
celestial orbs which circled round it believed to
be very far distant.   Tycho Brahé imagined that
the stars were not much more remote than the
planets.   Epicurus thought the stars were small
crystal mirrors in the sky which reflected the solar
rays, and the Venerable Bede remarked that they
needed assistance from the Sun's light in order to
render them more luminous.

The adoption of the Ptolemaic system by Milton afforded greater scope for the exercise of his imaginative powers, and enabled him to bring within the mental grasp of his readers a conception of the universe which was not lost in the immensity associated with the Copernican view of things. Besides, it also furnished him with a distinctly defined basis upon which to erect the superstructure of his poem. Above the circumscribed universe was Heaven or the Empyrean; underneath it was Chaos, from which it had been reclaimed, and in the lowest depth of which Milton located the infernal world called Hell. These four regions embraced universal space; and in the elaboration of his great epic Milton relied upon his imaginative genius, his brilliant scholarship, his vast erudition, and the divine inspiration of the heavenly muse. With these, aided by the power and vigour of his intellect, he was enabled to produce a cosmical epic that surpassed all previous efforts of a similar kind, and which still remains without a parallel.

One of the distinguishing features of Milton's mind was his wonderful imagination, and in its exercise he beheld those sublime celestial and terrestrial visions on which he reared fabrics of splendour and beauty, described in harmonious numbers with the fervid eloquence and charm of a true poet. An example of the loftiness and originality of his imagination is afforded us in his description of the Creation, the main facts of which he derived from the first two chapters of Genesis, and upon these he elaborated in full and striking detail his magnificent conception of the efforts of Divine Might, which in six successive creative acts called into existence the universe and all that it contains. The rising of the Earth out of Chaos; the creation of light and of the orbs of the firmament; the joyfulness associated with the onward career of the new-born Sun; the subdued illumination of the full-orbed Moon, and the thousand thousand stars that spangle the nocturnal sky—all these afforded Milton a rich field in which his imagination luxuriated, and in the description of which he found subject-matter worthy of his gifted intellect.

Milton gives an ampler and more detailed description of the new universe in his narration of Satan's journey through space in search of this world, and brings more vividly before the imagination of his readers the glories of the celestial regions. The fiend, having emerged from the dark abyss of Chaos into a region of light, first beheld the new creation from such a distance that to his view it appeared as a star suspended by a golden chain from the Empyrean. This stellar conception of the poet's harmonised with the views of the Ptolemaists, who believed that the universe was of limited extent, and though its dimensions were vast beyond comprehension, it was, nevertheless, enclosed by the tenth sphere or Primum Mobile. It was on the surface of this sphere that Satan alighted, and over which he wandered, until attracted by a beam of light that appeared through an opening at its zenith, where, by means of a stair or ladder, communication was maintained between the new universe and Heaven above. Hither the undaunted fiend hied, and, standing on the lower steps of this structure, momentarily paused to gaze upon the glorious sight which burst upon his view before directing his flight down into the newly created universe. Milton then describes his progress through the stellar regions, his landing in the Sun and what he saw there, and the termination of his journey when he descends from the ecliptic down to the Earth. In doing so the poet gives a wonderfully beautiful description of the starry universe, of the Sun, Moon, and Earth (Book III. 540–742), enhanced and adorned with his own poetic imaginings derived from fable, philosophy, and science.

Milton makes more frequent allusion to the Sun than to any of the other orbs of the firmament. This we should expect : the poet always gives the orb the precedence which is his due, and never fails, when the occasion requires it, to surround him with the ' surpassing glory ' which marks his pre-eminence above all other occupants of the sky. The Moon, his consort—peerless in the subdued effulgence of her borrowed light ; the beautiful star of evening, Hesperus ; the sidereal heavens with their untold glories ; the Galaxy, overpowering in the magnificence of its clouds and streams of stars. —all these have their beauties and charms mirrored in the pages of this remarkable poem.

Z

That the observation of the celestial orbs, their phases, and the varied phenomena which occur as a consequence of their motions, were to Milton an unfailing source of enjoyment and of meditative delight, is evident from the frequency with which he alludes to them. The following lines also testify to this :—

> For wonderful indeed are all his works,
> Pleasant to know, and worthiest to be all
> Had in remembrance always with delight!
> But what created mind can comprehend
> Their number, or the wisdom infinite
> That brought them forth, but hid their causes deep?
>
> iii. 703–708.

It is very pleasant, as Milton says, to

> sit and rightly spell
> Of every star that heaven doth show.

It is also pleasant to know the astronomy of his 'Paradise Lost,' and to linger over the delightful and harmonious utterances associated with the sublimest of sciences, expressed in the melodious language of England's greatest epic poet.